W0107057

X-RAY AND ELECTRON PROBE ANALYSIS IN BIOMEDICAL RESEARCH

PROGRESS IN
ANALYTICAL CHEMISTRY
Based upon the Eastern Analytical Symposia

Series Editors:

Ivor L. Simmons
M&T Chemicals, Inc., Rahway, N. J.

and Paul Lublin
General Telephone and Electronics Laboratories, New York, N.Y.

Volume 1

H. van Olphen and W. Parrish
X-RAY AND ELECTRON METHODS OF ANALYSIS
Selected papers from the 1966 Eastern Analytical Symposium

Volume 2

E. M. Murt and W. G. Guldner
PHYSICAL MEASUREMENT AND ANALYSIS OF THIN FILMS
Selected papers from the 1967 Eastern Analytical Symposium

Volume 3

K. M. Earle and A. J. Tousimis
X-RAY AND ELECTRON PROBE ANALYSIS
IN BIOMEDICAL RESEARCH
Selected papers from the 1967 Eastern Analytical Symposium

Volume 4

C. H. Orr and J. A. Norris
COMPUTERS IN ANALYTICAL CHEMISTRY
Selected papers from the 1968 Eastern Analytical Symposium

PROGRESS IN ANALYTICAL CHEMISTRY
VOLUME 3

X-RAY AND ELECTRON PROBE ANALYSIS IN BIOMEDICAL RESEARCH

Edited by K. M. Earle
Chief, Neuropathology Branch
Armed Forces Institute of Pathology
Washington, D. C.

and A. J. Tousimis
President, Biodynamics Research Corporation
Rockville, Maryland

ℚ SPRINGER SCIENCE+BUSINESS MEDIA, LLC

ISBN 978-1-4899-5909-6   ISBN 978-1-4899-5907-2 (eBook)
DOI 10.1007/978-1-4899-5907-2

Library of Congress Catalog Card Number 79-85212

© 1969 Springer Science+Business Media New York
Originally published by Plenum Press, New York
Softcover reprint of the hardcover 1st edition 1969

All rights reserved

No part of this publication may be reproduced in any
form without written permission from the publisher

PREFACE

This volume contains six papers from the two sessions on X-ray spectroscopy and electron probe analysis as related to biological materials presented at the Eastern Analytical Symposium, November 8–10, 1967.

While the impact of these methods has been primarily in such fields as metallurgy, there is increased interest in these techniques as applied to biomedical research.

Included in this volume are detailed descriptions of the special specimen handling requirements of biological materials, analytical procedures, and the operational parameters leading to the best results. This area of application has been relatively unexplored up to now. It is hoped that the publication of this volume will stimulate increased efforts in this direction.

We wish to thank the session chairmen who were responsible for the editing of the papers and without whose help this volume would never have been published.

PAUL LUBLIN
IVOR SIMMONS

CONTENTS

I. X-RAY SPECTROSCOPY IN FORENSIC MEDICINE—METAL POISONING

D. F. Craston

Office of Chief Medical Examiner
New York City
and Department of Forensic Medicine
New York University, School of Medicine

X-ray spectrographic analysis is a very useful method for evaluation of certain element concentrations in biological tissues for purposes of clinical and toxicological investigation. The organic matrix prevents presentation of samples for analysis in their natural state; but after suitable isolation and in some cases preconcentration data of high accuracy and reproducibility are obtained with X-ray spectroscopy.

Methodology appropriate for clinical and toxicologic analysis with emphasis on metal poisonings will be presented, compared with other available methods and critically evaluated. The use of X-ray diffraction in forensic medicine and criminology will be discussed and documented with cases.

1. FORENSIC MEDICINE

1.1. Scope

Forensic medicine is that branch of medical science which investigates a human organism in order to establish, if and how it was involved in some violent act which is in conflict with the existing legal code. In a majority of cases a clinician and a pathologist may find the answers by examining the body and its tissues grossly and microscopically. Yet a great number of cases—accidental, homicidal, or suicidal—require the study by a toxicologist. Since the number of drugs and chemicals available to the public has increased considerably in the last two decades, the complexity of the chemical investigation has grown proportionally.

1.2. Methods of Analysis

A toxicological analysis follows one of the accepted schemes by which first large classes of drugs or chemicals are isolated. Narrowing the path stepwise by isolation and purification, the chemist hopes to identify in the final step one drug or a few of its derivatives as the causative agent.

The toxicology laboratory work uses all means of analytical chemistry from a simple distillation or titration up to the methodology of physical microdetermination which is now possible with the advanced design of laboratory apparatus.

X-rays have been exploited to establish new microanalytical methods, revealing both the submicroscopic structure and the chemical micro-composition of the specimen. There has not been a systematic application of these methods to problems in forensic medicine. But isolated cases are reported and methods are worked out for the use of microradiography, X-ray microscopy, X-ray crystallography, and X-ray absorption for analysis of tissue samples ([1-13]). More often do we come across references of the use of emission X-ray spectrography and X-ray diffraction studies applied to biomedical problems ([14-20]).

2. METALS AND THEIR TOXICITY

The bulk of living matter is formed by organic material of low Z-number elements up to number 20 (Ca) with only Fe_{26} and Zn_{30} above 20 being present in any significant amounts. The other elements of higher Z-number are present in traces only, or they are completely foreign to the body.

The body maintains a fine electrolytic balance, confining the physio-logically significant elements to constant levels within a relatively narrow range. Their deficiency or intake beyond the means of their physiological control leads to disturbance of the metabolic processes to a point of failure and possible death. Likewise, the introduction of foreign elements into the living organism, for which it has no equipment to handle, causes a dis-turbance proportional to the quantity introduced and to its relative "toxicity." Heavy metals are very toxic since they cause precipitation of the proteins of the protoplasm in which the metabolic processes are carried out. But some light metals, e.g., arsenic, are equally aggressive, acting as "respiratory poisons" inactivating the enzymes involved in the oxidative metabolic cycle.

Many factors affect the toxicity of a metal, such as the solubility of its compound. Thus, many organic insoluble salts of mercury or arsenic may be used as medication while their inorganic salts are highly toxic. Inorganic barium sulfate may be introduced into the digestive system for purposes of X-ray investigation without any harmful effects, though its soluble salts are highly toxic. While the action is usually widespread, there are target organs which are more affected by a specific metal and it depends on their level of integration within the living organism to reflect the toxicity of the metal. The rate of dissemination through the body is regulated also by the portal of entry, and that will reflect the effect of the toxic elements. Last but not least is the efficiency of the detoxificating mechanisms present in the body and its general condition of health or disease.

This demonstrates the wide range of metal poisons and the factors involved in the relative toxicity, showing that it is very important to have on hand a method of scanning with high sensitivity the analyzed tissue for the presence of the greatest possible number of elements.

3. X-RAY EMISSION SPECTROSCOPY IN METAL POISON DETERMINATION

X-ray spectrography has been used successfully as an analytical tool in many branches of industry. Only during the last ten years has it been applied to any extent in analysis of biomedical samples. For a toxicological laboratory it has a special appeal, because it has properties which may be summarized under the following headings.

1. *The instrument covers a wide range of elements.* The practically useful range of elements to be determined both qualitatively and quanti- tatively by X-ray emission spectrography is from Z-number 13 (aluminum) upward throughout the periodic table of elements. Up to element Z-number 56 (barium), the K-lines provide a strong analytical signal with a 50 kV instrument; L-lines are used for elements above number 56. With a 100 kV X-ray source one can reach the K-line of bismuth ($Z = 83$). The range of Z-numbers 22–56 is quoted as optimal with the today available equipment.

2. *The sensitivity of the instrument* goes to submicrogram quantities of pure elements. In general, it may be stated that 0.1–3 ppm quantities are the lower limit for determination of elements in an inorganic matrix.

3. *The size of the sample and its state.* The nature of the sample may vary from solid to liquid to gaseous and appropriate cells to introduce such samples into the instrument are available. The quantity of the sample required depends on its state, its purity, and the nature of the matrix in which it is embedded; but, in general, it can be stated in terms of micro- grams and microliters for a good quantitative determination.

4. *The method is nondestructive to the sample.* For all practical pur- poses X-rays do not affect the chemical nature of the sample, and after it has been analyzed by X-ray it may be used for another confirmatory analysis or it may be stored for later reference.

5. *Some samples require no preparation whatsoever.* Samples containing strictly inorganic material within the limits of sensitivity of the instrument may be used for analysis without any preparation. Preconcentration may be required where the ratio of element to matrix is too large.

6. *The spectrum is simple.* Since one deals with energies representing the *K* or *L* shells in the atomic structure, the spectrum does not suffer by the multitude of spectral lines occuring in emission spectra produced by other means of excitation (electric arc, flame, etc.) which involve the more peripheral electron transitions. The overlap of spectral lines is minimal. The resolution is very good with proper collimation and may be further improved with the use of pulse height analysis.

7. *Excellent reproducibility.* The steadiness of the energy source and the mode of excitation is continuous which result in an uniform spectrum, the intensity of which may be regulated and reproduced from one instance to another. Also, the background components and the matrix effect show this steadiness. The standard deviation between individual determinations averages 0.7–1.2%.

8. *Compactness of the X-ray instrumentation.* The component parts of an X-ray emission spectrograph take up little space and require no more than a power and cooling water supply. Also, the auxiliary chemical glassware and chemical equipment is reduced to a minimum.

9. *Permanence of record.* From the point of view of legal documentation, the instrument coupled with a chart recorder provides a permanent record which also can be calibrated.

3.1. Specimen Handling

In relation to the body under investigation, the samples are classified as external and internal. The former are substances suspected to be involved in the case of metal poisoning (evidence). These may be relatively pure chemicals, drugs or household preparations gathered at the scene. They may be various objects like toys, chips of plaster or paint, contaminated food or drink, or other solutions prepared as domestic remedies, e.g., abortifaciens.

The internal samples are removed from the body's cavities such as the respiratory passages, various compartments of the alimentary tract, and the genito-urinary tract, or they are tissues of the body removed by biopsy or at the autopsy and submitted for analysis. The complexity of composition of the sample varies from an inorganic compound to an organic tissue of an organ of a body which might have been submerged into water, is decomposed and contaminated.

The steps of isolation, purification, and concentration of the toxic agent form the major part of the toxicologist's work. The identification is a relatively simple instrumental step, but it determines to what degree the former three procedures have to be carried out. In this respect, the X-ray emission spectrographic analysis is less demanding than other analytical methods. Purification is reduced to removal of the organic matrix or the water content if the metal levels are very low, and concentration is usually not required due to the high sensitivity of the instrument. The removal of organic matrix may be done in two ways.

1. *Ashing by heat*. The sample is first dehydrated at 110°C for a period of 24–48 hr and then ashed at temperatures of 420–430°C leaving a carbonaceous ash residue. The low ashing temperature prevents loss of some volatile chlorides. This method is unsuitable for compounds volatile in this range of temperature, such as mercury, antimony, and arsenic compounds.

These require preparation by:

2. *Wet ashing*, i.e., oxidative degradation of the organic matrix at temperatures of 100–200°C, depending on the nature of the sample in concentrated nitric, perchloric, and sulfuric acids. The acidity of the digest is neutralized by ammonia. Aliquots of the solution are either plated on metal planchettes or on filter paper. If concentration is required, it may be done by evaporation, complexation, or chromatography.

3.2. Standardization

For simple, inorganic liquid samples, aqueous solutions of the metal standards are plated on the same support as the unknown. Based on the counting rate, a standard curve is established and the results are plotted on it.

For organic samples either wet or dry ashed, the standards are prepared by adding three or four known low concentrations of the standard solution to an accurately weighed sample of the tissue. These standards are processed in exactly the same way as the unknowns. Again a curve is established, and the value of the unknown metal is found by interpolation on the curve below the lowest standard value. The curves are linear for a wide range of concentration. Duplicate standards show standard deviations of 0.3–0.6%; duplicate samples of organic material processed by wet or dry ashing show 0.7–1.2%.

3.3. Application

There are three light elements which are of interest to a toxicologist and which cannot be investigated by X-ray emission spectrography, namely, beryllium, boron, and fluorine. The elements of $Z = 10–16$ are better quantitated by other methods, but for qualitative analysis the X-ray is quite suitable. The remainder of the periodic table of elements is suitable for both qualitative and quantitative determination, and in Table I we show the elements and the criteria by which we determined them by X-ray emission spectrography in our laboratory. The results were compared with results obtained simultaneously by other methods (emission arc spectroscopy,

TABLE 1

Parameters Used in X-ray Determination of Some Elements in Toxicologic Samples

Element	Radiation	Crystal	Path	Counter	Support
P	$K\alpha$	ADP/EDDT	VAC.	[a]GFP	[c]FP
S	$K\alpha$	NaCl	VAC.	GFP	[d]AlP
Cl	$K\alpha$	NaCl	VAC.	GFP	AlP
K	$K\alpha$	LiF	VAC.	GFP	AlP
Ca	$K\alpha$	LiF	VAC.	GFP	AlP
Ti	$K\alpha$	LiF	VAC.	GFP	FP
V	$K\beta$	LiF	VAC.	GFP	FP
Cr	$K\alpha$	LiF	VAC.	GFP	FP
Mn	$K\alpha$	LiF	AIR	[b]SCI	FP
Fe	$K\alpha$	Same conditions for all subsequent elements			
Co	$K\beta$				
Ni	$K\beta$				
Cu	$K\alpha$				
Zn	$K\alpha$				
As	$K\alpha$				
Se	$K\beta$				
Br	$K\alpha$				
Sr	$K\alpha$				
Ag	$K\alpha$				
Cd	$K\alpha$				
Sn	$K\alpha$				
Sb	$K\alpha$				
Te	$K\alpha$				
I	$L\alpha$				
Ba	$K\alpha$				
Au	$L\beta$				
Hg	$L\beta$				
Tl	$L\alpha$				
Pb	$L\alpha$				
Bi	$L\alpha$				

The instrument used - Norelco 60 KB Vacuum Spectrograph
[a] GFP—Gas flow proportional counter
[b] SCI—Scintillation counter
[c] FP—Filter paper discs ($\frac{1}{2}$in. dia. Whatman #40)
[d] AlP—Aluminum planchette ($\frac{1}{2}$in. dia. sample well in Reynolds Wrap heavy duty)

flame spectroscopy, atomic absorption, and complexometric methods) and showed an average accuracy within 2.8%.

3.4. Summary

Emission X-ray spectrography is very well suited for a toxicologic laboratory. It is accurate and reproducible, and the sample is available for further analysis—this is important from the point of legal evaluation of a toxicologic report. The method is quick and requires little of accessory equipment.

The cost is in the high bracket of spectrographic instrumentation, but the versatility and ease of operation do justify this expense. As methodology is worked out in future years, exploitation of this method in the field of forensic medicine will be even more rewarding.

REFERENCES

1. A. Engstrom, Quantitative Microchemical and Histochemical Analysis of Elements by X-rays, *Nature* **158**, 664 (1946).
2. A. Engstrom, Microradiography in Biological and Medical Research, *Medica Mundi* **1**, 119 (1955).
3. R. Amprino, and A. Engstrom, Studies on X-ray Absorption and Diffraction of Bone Tissue, *Acta Anat.* **15**, 1 (1952).
4. C. Lagergren, Biophysical Investigations of Urinary Calculi: An X-ray Crystallographic and Microradiographic study, Thesis, *Acta Radiol., Suppl.* **133** (1956).
5. K. A. Omnell, B. Lindstrom, F. C. Hoh, and E. Hammerlund-Essler, Method for Non-destructive Determination of Inorganic and Organic Material in Mineralized Tissues, *Acta Radiol.* **54**, 209 (1960).
6. A. Engstrom, and B. Lindstrom, A Method for the Determination of the Mass of Extremely Small Biological Objects, *Biochim. Biophys. Acta* **4**, 351 (1950).
7. F. C. Hoh, and B. Lindstrom, On the Theory of Quantitative Microradiography in Biology, *J. Ultrastructure Research* **2**, 512 (1959).
8. B. Lindstrom, Roentgen Absorption Spectrophotometry in Quantitative Cytochemistry, *Acta Radiol., Suppl* **125** (1955).
9. A. Engstrom, Quantitative Micro- and Histochemical Elementary Analysis by Roentgen Absorption Spectrography, Thesis, *Acta Radiol., Suppl.* **63** (1946).
10. A. Engstrom, and J. B. Finean, Micro X-ray Diffraction in Histochemistry, *Exptl. Cell Research* **4**, 484 (1953).
11. V. E. Cosslett, A. Engstrom, and H. H. Pattee, *X-Ray Microscopy and Microradiography*, Academic Press, New York (1957).
12. A. Engstrom, V. E. Cosslett, and H. H. Pattee, *X-Ray Microscopy and X-Ray Microanalysis*, Elsevier Publishing Co., Amsterdam (1960).
13. H. H. Pattee, V. E. Cosslett, and A. Engstrom, *X-Ray Optics and X-Ray Microanalysis*, Academic Press, New York (1963).
14. L. S. Birks, *X-Ray Spectrochemical Analysis*, Interscience Publishers, Inc., New York (1959).
15. G. L. Clark, *The Encyclopedia of Spectroscopy*, Reinhold Publishing Corp., New York (1960).
16. H. A. Liebhafsky, H. G. Pfeiffer, E. H. Winslow, and P. D. Zemany, *X-Ray Absorption and Emission in Analytical Chemistry*, John Wiley and Sons, Inc., New York (1960).
17. G. H. Morrison, *Trace Analysis* (Physical Methods), Interscience Publishers, Inc., New York (1965).
18. E. Jackwerth, and H. G. Kloppenburg "Untersuchungen zur quantitativen Auwsertung von Papierchromatogrammen in der Spurenanalyse durch Roentgenfluoroescenzspectroskopie," reprinted from *Anal. Chem.* **179** (3), 186 (1961).

19. J. Cholak, and D. M. Hubbard, Determination of Manganese in Air and Biological Material, *Reprinted from Am. Ind. Hyg. Assoc. J.*, Vol. 21, No. 5 (Oct. 1960).
20. J. C. Mathies, and P. K. Lund, X-Ray Spectroscopy in Biology and Medicine—III. Bromide (Total Bromine) in Human Blood Serum, Urine and Tissues, Reprinted from the *Norelco Reporter*, **Vol. VII**, No. 5 (Sept.–Oct. 1960).

II. ELECTRON PROBE ANALYSIS OF HUMAN LUNG TISSUES*

William G. Banfield

National Cancer Institute, Bethesda, Maryland
and
A. J. Tousimis, J. C. Hagerty, and Thomas R. Padden

Biodynamics Research Corporation, Rockville, Maryland

During electron-probe analysis of asbestos bodies and anthracotic pigment in tissue sections of human lung, X-ray spectrometer scans for most elements of the periodic table yielded an unexpected result. Titanium was found in high local concentrations associated with some asbestos bodies and not others; and, as expected, iron, silicon, and magnesium were also present. Further examination of both lung and hilar lymph node tissue showed that some particles with the usual light microscopic appearance of anthracotic pigment also contained titanium. The association and proportions of titanium, iron, and silicon in particles in the lung and hilar lymph nodes was not constant. An individual with lung and hilar lymph nodes containing high titanium had worked as a painter. Titanium dioxide paint pigment might have been the source in this particular case.

1. ELECTRON PROBE ANALYSIS OF PARAFFIN SECTIONS

1.1. Specimen Preparation Methods

Paraffin sections suitable for light microscopy were mounted on quartz, epoxy, or copper 1-in. diameter discs, the paraffin removed with xylene, and the sections coated with carbon before examination in the electron probe. It is recognized that carbon rods used for spectroscopy contain titanium impurities. However, in controls of carbon coated control discs only rarely was a particle containing titanium encountered.

1.2. Analysis of Asbestos Bodies

Analyses of single asbestos bodies in sections from human lung from cases of asbestosis was carred out with the electron probe. Asbestos is a

*Work supported by National Institutes of Health Contract No. PH-43-67-104.

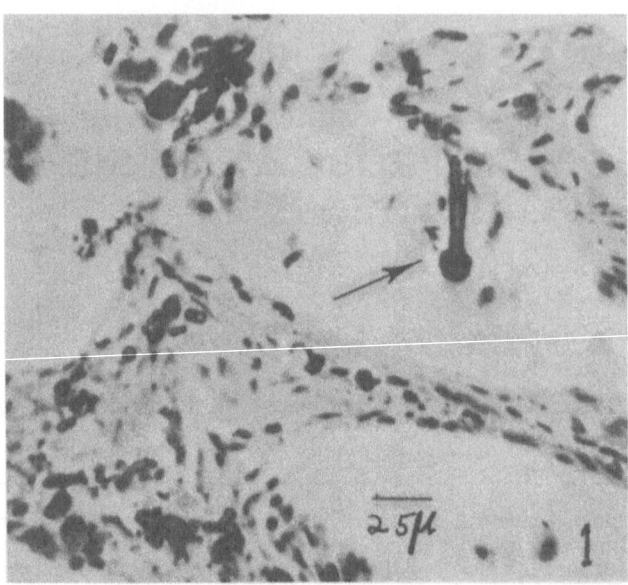

Fig. 1. Asbestos bodies. One at arrow. Light micrograph.
Paraffin section of lung is stained with hematoxylin and eosin.

mineral that occurs as fibers and it is only those less than 5 μ, usually less
than 1 μ in diameter, which reach the lung to do damage. The length of
the fibers is variable but may be as long as 50 or 60 μ. In the lung they often
become coated with an organic material containing iron and calcium,
forming structures known as *asbestos bodies*. These bodies are associated
with fibrosis which can become severe, impairing the function of the lung
and leading to the symptoms of asbestosis. In a high percentage of cases,
asbestosis is associated with carcinoma of the lung ([1]). So, the ability to
analyze these asbestos bodies and to establish by analysis what type of
asbestos is associated with them is both of practical and fundamental
importance. Figure 1 is a light micrograph of a section of lung tissue from
a case of asbestosis. The alveolar walls are thickened, contain much
fibrous connective tissue, and there are asbestos bodies within the walls.
The shape of the asbestos body projecting into the alveolus is typical. It is
club shaped with the head measuring about 12 μ in diameter, the shaft 5 μ,
and the entire body approximately 50 μ in length. In an X-ray spectrometer
scan of such a body in the electron probe there was found in addition to
the iron, calcium, silicon, and magnesium which were expected, a high
titanium peak. Other asbestos bodies were scanned, some contained
titanium and others did not.

1.3. Analysis of Anthracotic Pigment in Sections

Let us now consider inhalation of coal dust. Here again, if the particles are on the order of 1 μ in size, they are able to reach the lung and are picked up by cells called macrophages. When seen in the light microscope, the coal dust appears black and is called anthracotic pigment. Figure 2 is a section of lung from a city dweller which contains masses of macrophages filled with this black pigment. Figure 3 is a higher power in which the pigment can be seen distributed in the configuration of cells.

All the particles seen are probably not coal dust. The pathologist, however, cannot tell this so he calls anything that looks like the black pigment in Fig. 3, *anthracotic pigment*. Using the same tissue sections in which titanium was found associated with the asbestos bodies, an electron probe analysis for titanium was carried out on a number of macrophages filled with anthracotic pigment. Again, as with the individual asbestos bodies, titanium was found in some macrophages and not in others.

To find how widespread the association of titanium with anthracosis might be, lung tissue from 13 individuals who had as one of their diagnoses anthracosis was retrieved from the National Institutes of Health autopsy files. These observations are summarized and shown in Table I. Six of the

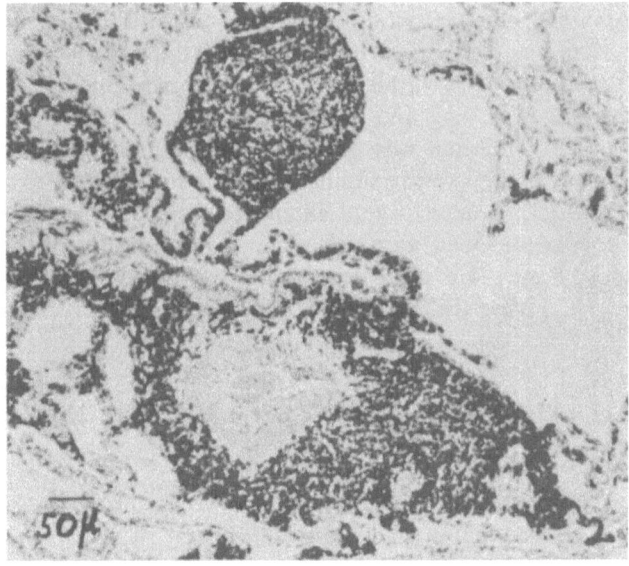

Fig. 2. Anthracosis. Cells filled with black material, "anthracotic pigment." Light micrograph. Paraffin section of lung is stained with hematoxylin and eosin.

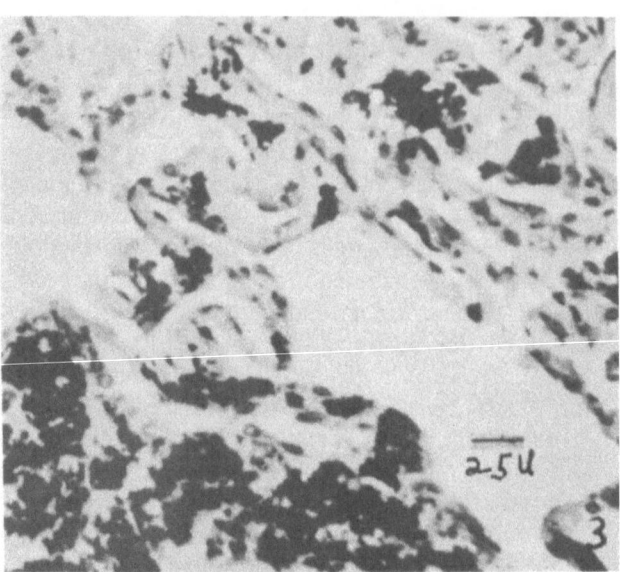

Fig. 3. Anthracotic pigment filling cells, macrophages. Light micrograph. Paraffin section of lung is stained with hematoxylin and eosin.

13 lungs examined contained titanium when only those areas with more than 0.1% titanium (calculated by first approximation) were considered positive. There was a wide variation in the number of positive to negative anthracotic areas in a lung and the content of titanium per area varied from 0.1 to 10%. Titanium was present in the hilar lymph nodes of the two cases in which they were examined. It was not present in local concentrations of 0.1% in other organs examined from one case in which the concentration per area and number of positive areas in the lung was relatively high. Figure 4 is a portion of an X-ray goniometer scan showing the titanium peaks when the electron beam was positioned on an anthracotic particle.

2. LOCALIZATION OF ELEMENTS IN TISSUE SECTIONS

2.1. Paraffin Sections

A major problem in electron probe analysis of tissue sections is the identification of the structures being analyzed.

Figure 5 is a light micrograph of macrophages containing anthracotic pigment in the lung. Figure 6 is a scanning electron micrograph (backscatter electrons) of the same area and Fig. 7 a similar picture at higher magnification. In the latter, the individual boundaries probably represent cells and

TABLE I

Titanium Occurrence in Anthracotic Pigment by Electron Probe Analysis

Patient no.	Organ examined	Number of particles found after diligent search		
		>1% Ti	1.0–0.1% Ti	<0.1% Ti
1	Lung	1	17	35
1	Hilar Lymph Node	19	30	12
2	Hilar Lymph Node	1	—	3
3	Lung	—	—	9
3	Lung	—	—	12
4	Lung	—	—	11
5	Lung	—	—	9
6	Lung	8	8	1
7	Lung	1	2	5
8	Lung	—	1	9
9	Lung	1	2	—
10	Lung	—	—	6
11	Lung	—	—	11
12	Lung	—	—	8
13	Lung	—	—	4
14*	Lung	10	3	1

*Other tissues examined: Uterus, Spleen, intestine, Thyroid, muscle, fat, ganglion, gall bladder, intestine, and stomach, all negative for titanium.

point for point it can be matched with the light micrograph. Figure 8 is a rate meter scan for Fe $K\alpha$ X-rays and Fig. 9 for Ti $K\alpha$ X-rays. The matching here of the microscopic picture of the tissue with the picture in terms of an element approaches cellular resolution in the analysis.

2.1.1. *Improved Instrumentation* (*Combined Scanning Electron Microscopy—Primary Backscattered Electrons—and X-ray Microanalysis*)

Figures 10–12 are examples of morphologic resolution obtained with improved instrumentation.* This consisted of adding a third lens, additional shielding to the electron optical system, and improved electronics.

2.2. 1-μ Sections of Methacrylated Embedded Tissue

Morphologic resolution can be further increased by using sections of tissue processed as for transmission electron microscopy. The tissue was fixed in glutaraldehyde, post fixed in veranol buffered osmium tetroxide, and embedded in methacrylate. One-micron sections were cut, mounted on quartz discs, the methacrylate removed with xylene, and the specimens coated with carbon in a high vacuum evaporator. Figure 13 is a light

*Materials Analysis Company, Palo Alto, Calif.

X-RAY GONIOMETER SCAN OF "ANTHRACOTIC" PARTICLE WITHIN LUNG MACROPHAGE

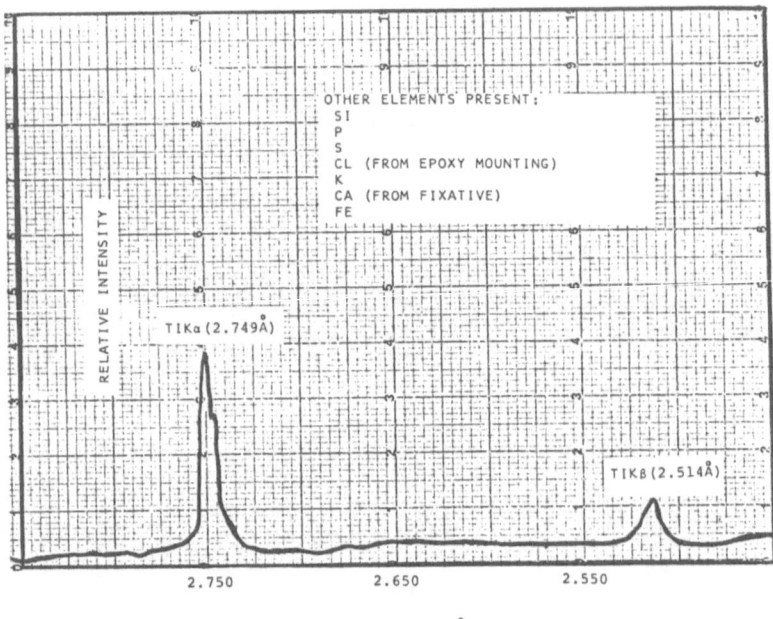

Fig. 4. X-ray spectrometer scan of asbestos body through titanium peaks.

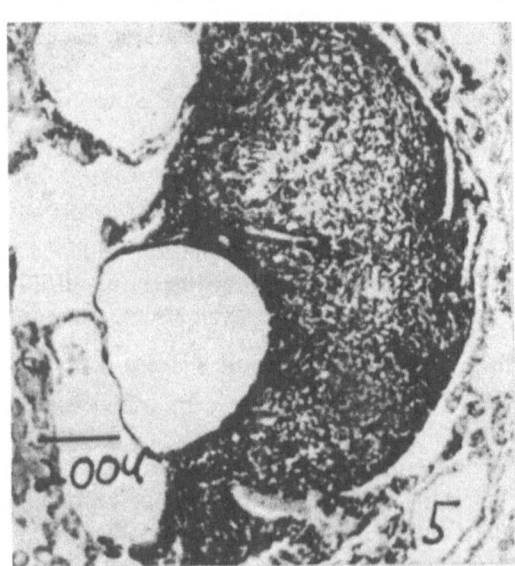

Fig. 5. Large focus of anthracotic pigment in lung with anthracosis. Light micrograph. Paraffin section.

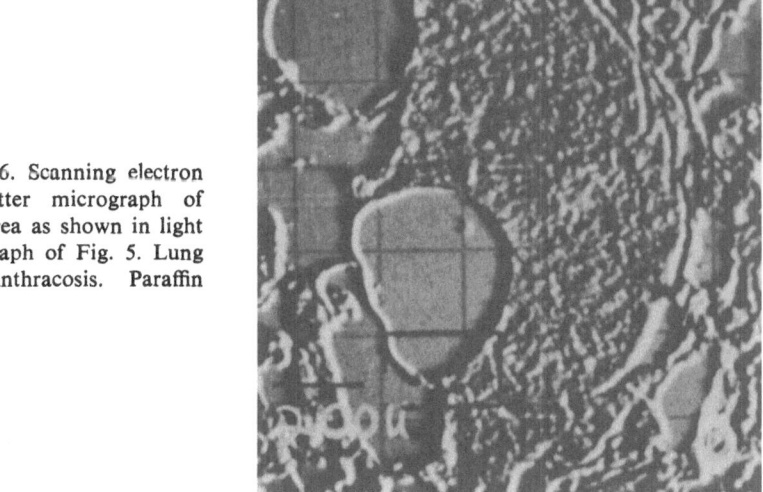

Figure 6. Scanning electron backscatter micrograph of same area as shown in light micrograph of Fig. 5. Lung with anthracosis. Paraffin section.

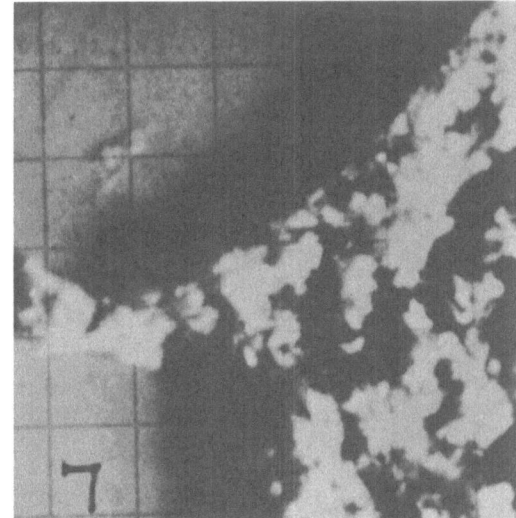

Fig. 7. Scanning electron micrograph using backscatter primary electrons of the area outlined in the square of Fig. 6. Lung with anthracosis. Paraffin section.

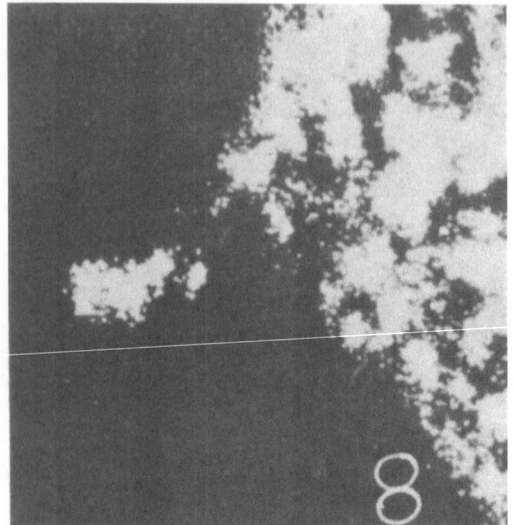

Fig. 8. Iron $K\alpha$ X-ray scan of area shown in Fig. 7. Lung with anthracosis. Paraffin section.

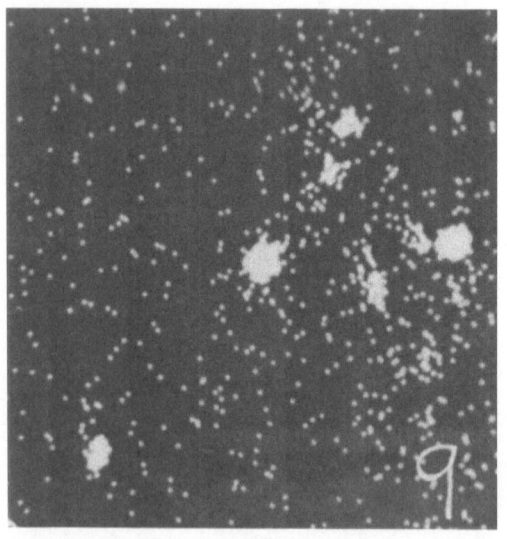

Fig. 9. Titanium $K\alpha$ X-ray scan of area shown in Fig. 7 slightly rotated. Lung with anthracosis. Paraffin section.

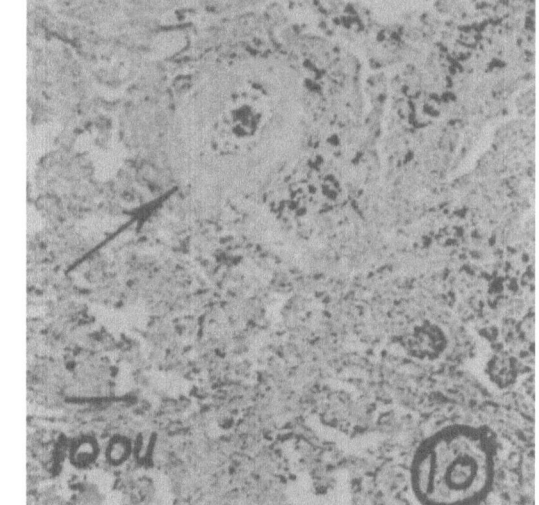

Fig. 10. Light micrograph. Lung with anthracosis. Paraffin section.

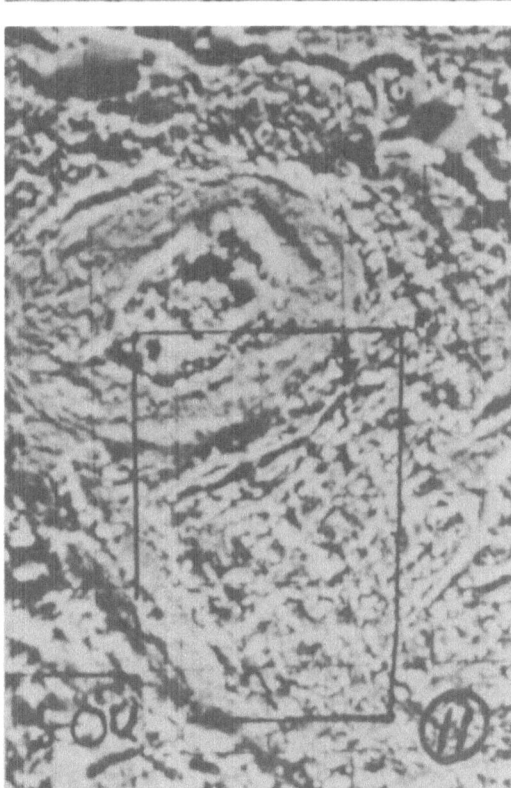

Fig. 11. Scanning electron micrograph (backscatter primary electrons) of blood vessel shown at arrow in Fig. 10. Lung with anthracosis. Paraffin section.

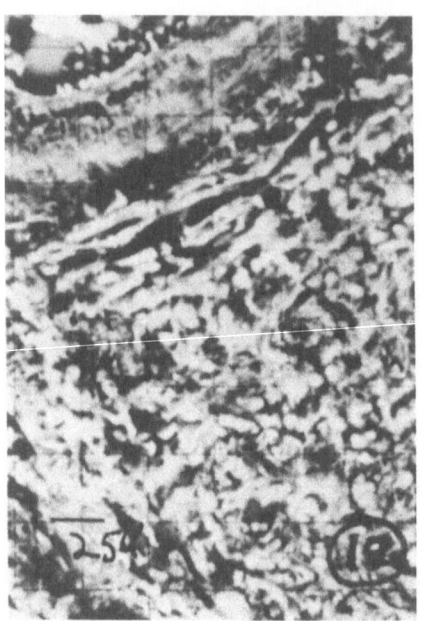

Fig. 12. Scanning electron micrograph (backscatter primary electrons) of area outlined in Fig. 11. Lung with anthracosis. Paraffin section.

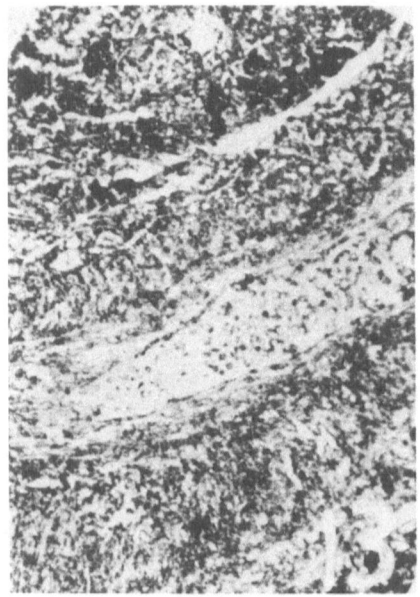

Fig. 13. Light micrograph. Lung with anthracosis. A 1 μ section of methacrylate embedded tissue is mounted on quartz disc and methacrylate removed with xylene.

micrograph of such a section, and Figs. 14–15 electron backscatter pictures of the same section at progressively higher magnification.

3. TRANSMISSION SCANNING ELECTRON MICROSCOPY*

Further morphologic resolution is desirable for certain applications, therefore, a new method of scanning tissue sections with the electron probe which has the potential of greatly increasing the resolution was tried. The tissue was embedded in epon or methacrylate as for transmission electron microscopy and sectioned with an ultramicrotome. The sections were 0.2 μ in thickness. They were mounted on 200-mesh parlodion film covered electron microscope grids and coated with carbon. They were placed in a holder in a combined scanning electron microscope electron probe assembly. Beneath the grid was a solid-state electron detector. The beam was scanned over the section, and the electrons transmitted through the section were registered by the detector which modulated an oscilloscope. An accelerating

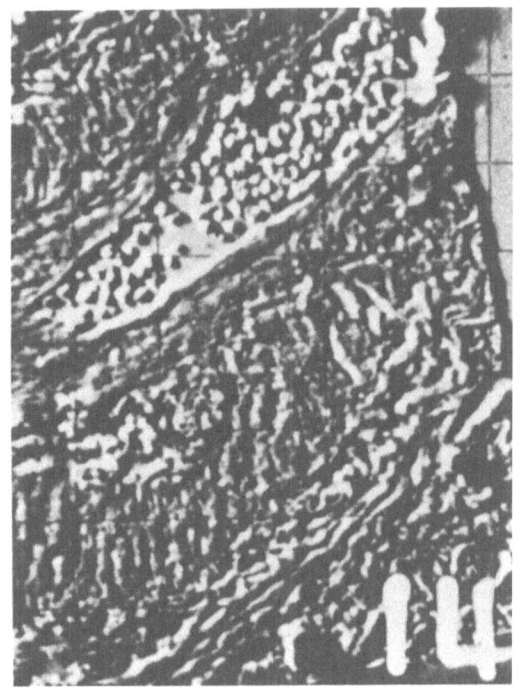

Fig. 14. Scanning electron micrograph (backscatter primary electrons). Lung with anthracosis. A 1 μ section of methacrylate embedded tissue mounted on quartz disc and methacrylate removed with xylene.

*This work was performed in collaboration with Nelson Yew of the Materials Analysis Co., Palo Alto, Calif.

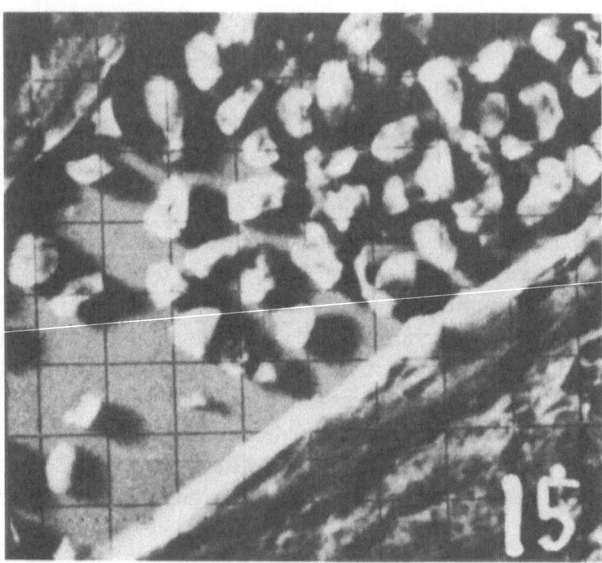

Fig. 15. Blood vessel seen in Fig. 14 containing red blood cells at high magnification. Lung with anthracosis. A 1 μ section of methacrylate embedded tissue is mounted on quartz disc and methacrylate removed with xylene.

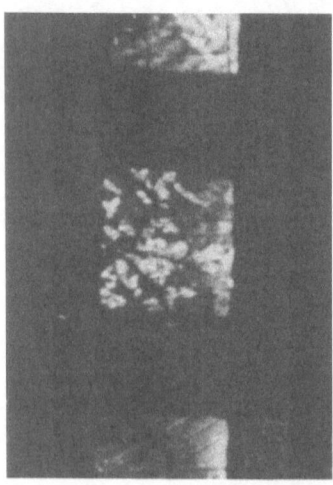

Fig. 16. Macrophages containing anthracotic pigment. Electron scanning transmission pictures of lung with anthracosis. A 0.2 μ section of epon embedded tissue is mounted on copper grid.

voltage of 4 kV and a current of 9×10^{-11} A was used. Figures 16 and 17 are low- and high-magnification scanning electron micrographs. The outlines of the macrophage, the nucleus, and the granules of anthracotic pigment within the cytoplasm are easily seen.

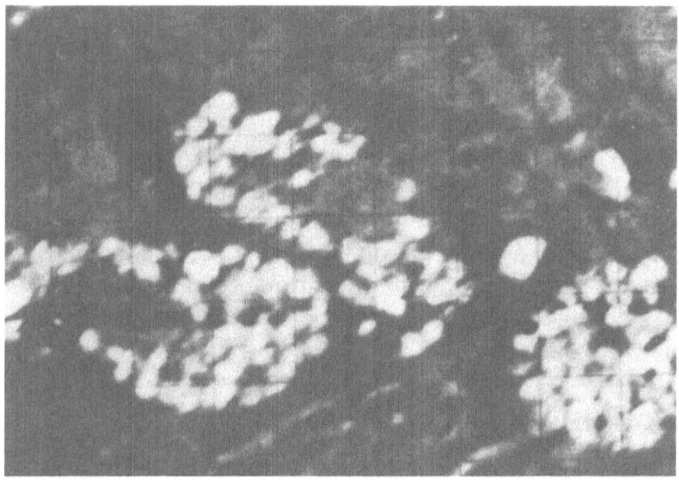

Fig. 17. Macrophages seen at arrow in Fig. 16. Electron scanning transmission pictures of lung with anthracosis. A 0.2 μ section of epon embedded tissue is mounted on copper grid.

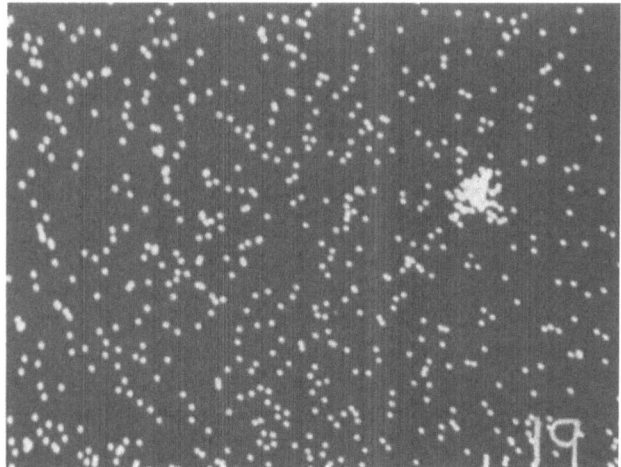

Fig. 18. Iron $K\alpha$ X-ray scan over macrophage at arrow in Fig. 16. Spot rich in iron also contained titanium.

3.1. X-ray Signal from Thin Sections

A significant X-ray signal using dispersive analysis can be obtained even from these 0.2 μ sections. Figure 18 is a scan for iron over one of the macrophages. Iron is present, and even though titanium could not be seen in a similar display, it could be detected when the electron beam was localized over the area of the iron signal.

3.2. Resolution

The resolution obtained in our first trials with scanning transmission was about 800 Å. With further refinement, much better scanning transmission resolution should be obtained with the electron probe thus making it possible to visualize morphology in the range of conventional transmission electron microscopy. In biology, most of the important tissue structures lie in this range, and in order to interpret results it is necessary to see what is being analyzed.

4. DISCUSSION

Considering again the titanium, the question was asked as to what state it was in. Although it was often associated with iron and silicon, there was no consistent ratio, and at times titanium occurred without iron and vice versa. Any attempt to find whether the titanium occurred as the oxide or the carbide would be frustrated by the presence of other particles in close association with the titanium containing particles.

Where did the titanium come from? It is well known that titanium oxide is a white pigment and is used in white paints and even in paints of other colors. The first patient in which we found titanium was a painter, which might explain the presence of titanium in this particular instance. Since there are asbestos paints, it may also explain why in this patient, who also had asbestosis, there was titanium associated with some asbestos bodies and not with others. However, titanium dioxide is often found associated with asbestos bodies isolated from lung and can give rise to pseudo-asbestos bodies ([2]). A patient with anthracosis was a domestic house worker and she had particles containing the highest concentration of titanium in her lungs. Another patient with anthracosis was a retired printer and another had been a store administrator. Analysis of soot collected by filtration of the air from three large cities did not show titanium. It has been reported that titanium occurs in most of the earth's crust in oxide concentrations of between 0.2 and 2.3 w % ([3]), so the possibility of it occurring in dust is good. Titanium was first reported in lung tissue and anthracotic lymph nodes in 1931 ([4]) and its distribution in the population unassociated with asbestosis extensively studied since then ([5]). It is considered inert and unlikely to have a harmful effect on the lungs ([6]).

Without the use of the electron probe elements which may occur in high local concentration might not be found, and to *localize* "trace elements" at the cellular level is difficult and in some cases impossible. We have then a new tool in the study of air pollution and human disease which allows us to uncover additional factors which may be important in our understanding of pathological processes.

REFERENCES

1. M. Telishi and A. I. Rubenstone, Pulmonary Asbestosis Associated with Primary Carcinoma of the Lung, Bronchial Adenoma and Adenocarcinoma of the Stroma, *Arch. Pathol.* **72**, 116–243 (1961).
2. N. Sundius and A. Bygden, Staubinhalt einer Asbestosislunge und die Beschaffenheit der sogenannter Asbestosiskörperchen, *Arch. Gewerbepathol. Gewerbehyg.* **8**, 26–70 (1937).
3. K. C. Condie, Composition of the Ancient North American Crust, *Science* **155**, 1013–1015 (1967).
4. K. F. Sheid, Über die Methodik der Darstellung und Bestimmung der in pneumonokoniotischen Geweben abgelagerten Staubes, *Beitr. z. pathol. Anat. u. allgem. Pathol.* **89**, 93–134 (1931).
5. H. A. Shroeder, J. J. Balassa, and H. Tipton, Abnormal Trace Metals in Man, Titanium, *J. Chronic. Dis.* **16**, 55–69 (1963).
6. H. Christie, R. J. Mackay, and A. M. Fisher, Pulmonary Effects of Inhalation of Titanium dioxide by rats, *Amer. Industr. Hyg. Assn. J.* **24**, 42–46 (1963).

III. X-RAY FLUORESCENT SPECTROS-COPY OF BRAIN TISSUE IN CASES OF PARKINSON'S DISEASE

Kenneth M. Earle

Chief, *Neuropathology Branch, Armed Forces Institute of Pathology*
Washington, D.C.

Histological studies were performed on 105 cases of Parkinson's disease in the files of the Armed Forces Institute of Pathology. The principal pathological changes will be reviewed.

In 11 cases elemental profiles of K, S, Cl, P, Ca, Fe, and Zn were determined on formalin fixed, dried pelletized brain tissue from various topographic areas. A relative increase of Fe and a decrease of K were observed. The possible significance of these findings will be discussed.

1. INTRODUCTION

In the vast majority of patients suffering from Parkinson's disease the cause is unknown. Viral encephalitis, inhaled manganese dust, certain drugs, trauma, neoplasms, cerebral hemorrhage, heavy metals, and arteriosclerosis have been suspected in specific cases, but the relationship of these agents, if any, to Parkinson's disease is obscure.

As part of an extensive study ([1]) of the pathology of Parkinson's disease based upon 513 documented cases in the files of the Armed Forces Institute of Pathology, we obtained sufficient formalin fixed brain tissue on 11 cases to do X-ray fluorescent spectroscopy. The details of this study and the over-all conclusions have been reported elsewhere. Our main goal was to determine if there were any major elemental shifts in the Parkinsonian brain tissue.

The purpose of this paper is to describe the uses and limitations of X-ray fluorescent spectroscopy in determining elemental profiles of biologic tissues, specifically human brain tissue, using ordinary commercially available equipment and simple dehydration and pelletizing as a means of sample preparation.

2. MATERIALS AND METHODS

The theories, technics, applications, and limitations of X-ray fluorescent spectroscopy can be found in various publications ([2-12]).

A diagram of the basic equipment is illustrated in Fig. 1. The X-rays from the tube are incident upon the sample that is inclined at a 45° angle. The elements within the sample emit secondary ("fluorescent") X-rays in addition to scattered radiation from the X-ray tube. Each element in the sample will emit its characteristic radiation if the voltage on the X-ray source is sufficiently high. For lighter elements, the $K\alpha$ line is ordinarily the best analytical line and the $L\alpha$ is usually the best line for the heavier elements under ordinary operating conditions. The fluorescent radiation is dispersed by the analyzing crystal. When Bragg's law ($n\lambda = 2d \sin \theta$)* is satisfied, the detector assembly will register the intensity of the fluorescent radiation and the goniometer will register the 2θ angle. The element is identified by solving Bragg's law or by referring to tables of 2θ angles for each element for each type of crystal of known $2d$ spacing ([13]).

We chose X-ray fluorescent spectroscopy because (a) It offers a wide range of analyses from atomic number 13 (Al) to 92 (U); (b) It is nondestructive, and analyses can be repeated over and over on a single small sample of less than 0.5 g; (c) The spectral pattern is relatively simple;

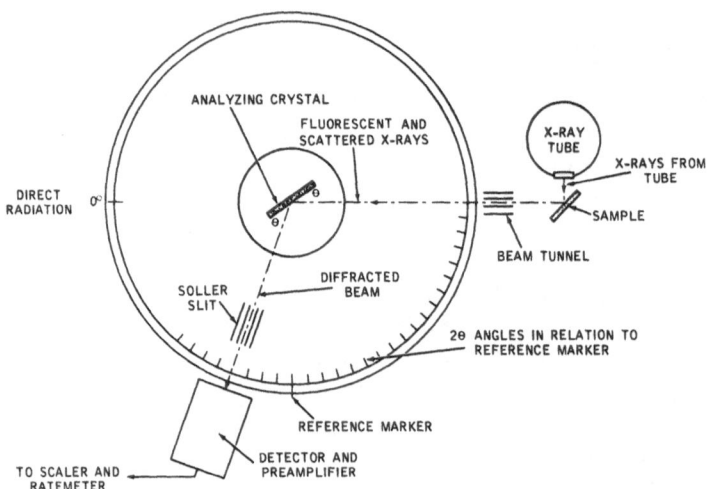

Fig. 1. Diagram of X-ray fluorescent spectrometer (see text). (AFIP Neg. 67-2741-1.)

*Where n is an integer, λ is the wavelength of the radiation, d is the interplanar spacing of the crystal that is cut parallel to a set of hkl planes, and θ is the angle of incidence and the angle of "reflection."

and (d) With ordinary equipment, the method is best suited for elements that exist in the order of 100 μg/g of dried pelletized sample, although elements can be detected at much lower levels with special technics. The lowered sensitivity is an advantage when searching for toxic levels of an element (in the order of 2 or more times the maximum range of normal) because it tends to lessen or eliminate the contribution of trace elements in the range of normal fluctuations. Likewise, it lowers the detection of trace contaminants that are inevitable in human tissues. In our experience, it does not detect low levels of trace contaminants in the water and formalin that are used in fixation. Of course, gross contamination would destroy the value of any spectroscopic analysis ([14]), but we have not found the contamination by fixatives to be very troublesome in X-ray spectroscopy, when compared with nonfixed material, in the range of sensitivity in which we are working; namely 100 μg/g or more in the usual case. In regard to the extent to which contamination should be avoided, we agree with DalCortivo and Cefola ([15]) that "The function of the toxicologist, as an analyst, is to achieve the highest accuracy and precision attainable, but as a toxicologist, his job is to establish presence or absence of intoxication in the shortest time and in as practicable a way as is possible."

After numerous experiments on sample preparation including desiccation, freeze drying, hot ashing, "cold" ashing, fusion in glass, chemical extraction, and liquid and dry samples, we have finally concentrated upon samples prepared as follows: Gray matter was separated from the white (or mixed if the samples were too small) by dissection with stainless steel instruments, weighed, sliced into pieces 1 × 1 cm and approximately 3-mm thick, desiccated overnight in silica dishes at 105° to 110°C in a hot air oven, ground for 10 to 29 min in a Spex shaker grinder in plastic vials that had tungsten carbide end caps and a tungsten carbide marble. Then, 500 mg of the dry powder were weighed carefully, placed in an ARL press, backed with 1 gm of chromatographic-grade cellulose powder, and pressed into a pellet of 1-in. diameter at 60,000 psi for 3 min. The pellets were wrapped in Saran wrap and stored in pill boxes. We have not found it necessary to keep them in a desiccator for reasonably reproducible results. (Our laboratory has evenly controlled central air-conditioning with no windows.)

The samples were placed in a General Electric XRD-6 X-ray fluorescent spectrograph as diagramed in Fig. 1. Each of the samples in this study was scanned qualitatively. In the first run for the lighter elements, an EA75 chromium target X-ray tube and PET (penta erythritol tetanitrate) ($2d = 8.742$ Å) crystal were used with the following operating conditions: 50 KVP, 40 ma, constant-potential, flow-proportional counter using P-10 gas, helium path; E 5 V; ΔE off; maximum amplification voltage on the counter adjusted so that Al pulses measured 10 V (hence, all of the other elements would be higher voltage in relation to the Al setting), log scale, chart speed 4° 2θ/min. The results of the qualitative scan on a sample of Parkinsonian brain tissue using a PET crystal and Cr (Chromium) radiation are shown in Fig. 2. A run from 148° 2θ to 10° 2θ requires a little over 30 min. By starting at the highest angle and sweeping to the lower angles, the graph reads from left to right as lower to higher, and the electrical cables descend more easily than they ascend.

The sample was then rerun using a Mo (molybdenum) target X-ray tube and a LiFl crystal ($2d = 4.028$ Å) for the heavier elements. The same operating conditions were used except that the voltage on the counter was adjusted so that the voltage of K (potassium) $K\alpha$ peaks was 10 V and all of the other elements would be accordingly higher. The pulse heights and intensities were monitored by an oscilloscope connected to the amplifier output circuit and by the ratemeter and scaler circuits. By operating at maximum amplification and the lowest possible voltage for a pulse height of 10 V for the lowest atomic number element of interest, we gained stability of operation. The results of a qualitative scan of a sample run with a Mo tube and LiFl crystal are shown in Fig. 3.

In special cases, other crystals (NaCl, EDT, KAP, silicon, and topaz) and a W (tungsten) target X-ray tube were used, but we found that the Mo and Cr X-ray tubes and the LiFl and PET crystals covered the range of elements that we found in brain tissues, and the other crystals resulted only in improvement of intensity of certain specific elements.

In both of the qualitative scans, the chart paper was rewound back to the starting point, and the technic was repeated with a cellulose blank and normal brain tissue as shown in the copied triple traces in Figs. 2 and 3, or individual runs were recorded and compared.

After the qualitative scans were completed, the number of counts per second for each major element was determined by setting the goniometer at the proper angle, using the pulse height selector with E 5 V, ΔE 10 V,* and bringing the pulse height of the element to be determined into the window by adjusting the voltage on the counter until the pulse of interest measured a pulse height of 10 V. The counts per second were compared with a cellulose blank and normal brain tissue for each element.

Absorption and enhancement effects were observed with analyses of the lighter elements when the lighter elements were added to normal brain standards or to cellulose, but these effects were less marked than was anticipated. In ashed samples the effect seemed to be greater than in the

---

*E is the baseline voltage and ΔE the voltage range in the "window" above the baseline voltage.

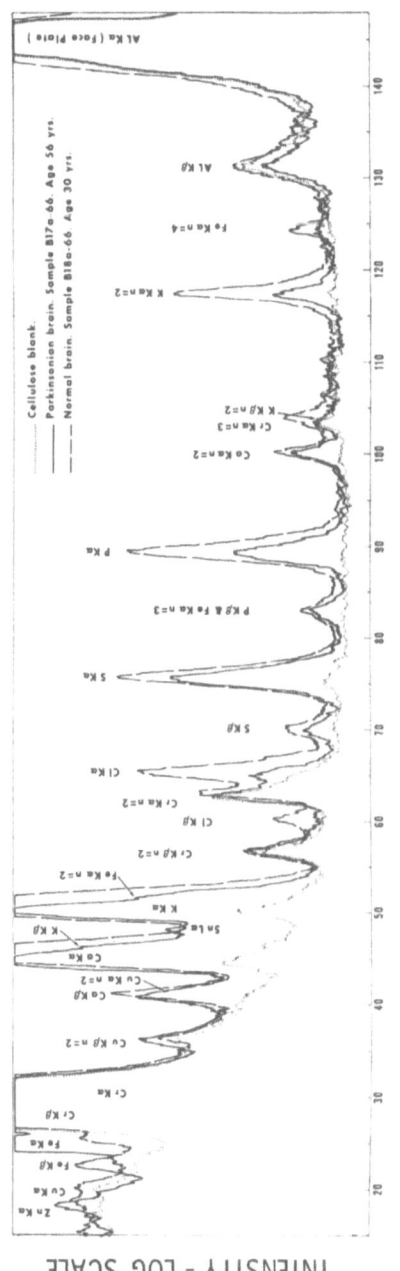

Fig. 2. Composite trace of 2θ angles (abscissa) *vs.* log scale of the intensity (ordinate) at maximum range for the GE XRD6 rate-meter for cellulose blank, Parkinsonian brain tissue, and on normal brain tissue. Note the high peaks of iron and low potassium in the Parkinsonian brain. Small amounts of Ca, K, and S are detected in the cellulose blank. The Cr, Fe, and Cu peaks in the cellulose blank are scattered radiation and contaminants from the X-ray tube. The Al peaks at the right of the scale are from the Al face plate of the sample holder. Chromium radiation; PET crystal; helium path; flow-proportional counter; 4° 2θ/min chart speed. (AFIP Neg. 67-2741-2).

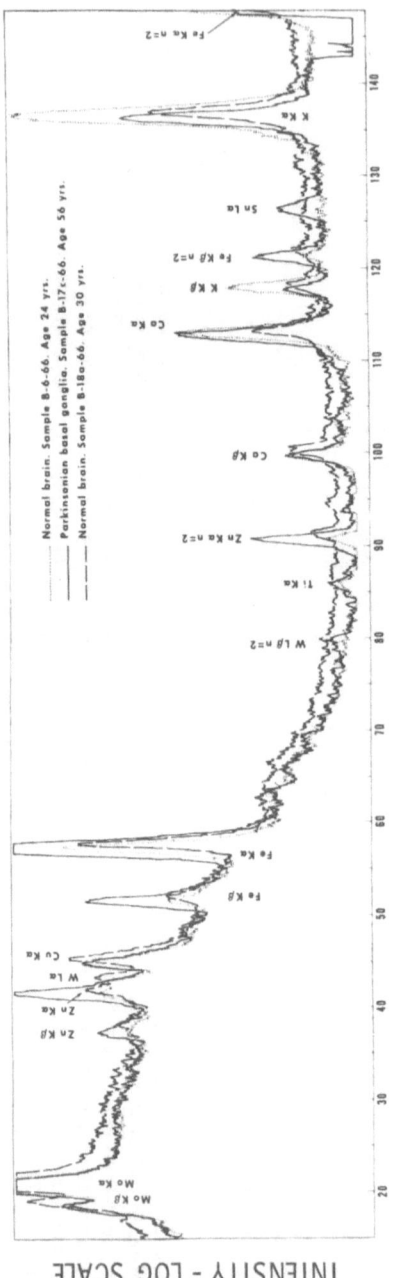

Fig. 3. Composite trace of 2θ angles (abscissa) *vs.* log scale of the intensity (ordinate) at maximum range for the GE XRD6 rate-meter for 2 specimens of tissue from normal brains to show normal fluctuations and 1 for Parkinsonian brain tissue using Mo radiation; flow-proportional counter; LiFl crystal; air path; $4°$ θ/min chart speed. Note the high Fe in the Parkinsonian and the variability of K in normal and Parkinsonian brain tissue caused, in part, by shift of this element into the formalin fixative. The peaks of Sn and Ti are contaminants. Some of the Fe, Cu, and W are scattered contaminants from the X-ray tube. The Mo peaks are scattered radiation of the tube-target element. (AFIP Neg. 67-2741-3).

dried samples due, we believe, to the dilution effects of the retained lipids in the brain samples. Of course, dried-brain samples yielded less intensity of fluorescent radiation than ashed samples, but they lost fewer elements of interest, and the matrix effect was lessened. The S and Cl were markedly reduced or lost by ashing, and such elements as Hg were completely lost by ashing but partly retained by drying at 105°C overnight.

Uncontaminated, unfixed, fresh, or fresh-frozen specimens will be necessary before accurate semiquantitative analyses can be undertaken. Accurate standards must be made and calibration curves constructed. We have prepared many such standards, but we are continuing to check stability before being assured of accuracy. Because no normal "standard" brain tissue exists, we are attempting to make reasonably reliable standards by using cellulose as a base, since this material scattered X-rays with intensity similar to dried brain tissues when pelletized at equal pressures for equal times.

3. OBSERVATIONS AND CONCLUSIONS

The X-ray fluorescent spectrographic analyses revealed good peak-to-background ratio of the following elements in both gray and white matter of Parkinsonian and normal brain samples: K, S, P, Cl, Ca, Fe, and Zn. Poor peak-to-background ratios were recorded for Cu. In the Parkinsonian brain tissues, the Fe was consistently increased by a factor of 2 or more above normal, and the K was consistently decreased; Zn was found consistently and cannot be considered a trace element in brain by usual definitions. The other elements showed no consistent shift, though moderate fluctuations were seen. One case showed a definite Br peak. Since bromides may accumulate in tissue, we suspect, but could not prove, that this patient had been on bromide medications prior to death. The analyses for Cu were hampered by contamination from the X-ray tubes. If major amounts of Cu existed in these samples, however, we could have detected it by comparison with the cellulose blank. The Fe is a contaminating radiation from most X-ray tubes, but the iron in brain tissue is high enough that the contaminant can be cancelled out by subtraction or by the use of Ti filters. The decrease of K was consistent and more marked than most of the samples of non-Parkinson brain tissues of similar age, but the result is open to question since K will shift readily into formalin solutions. We found that this shift of K was very marked when we compared samples of fresh brain tissue before and after formalin fixation.

The marked increase of Fe in the formalin-fixed Parkinsonian brain tissue when compared with controls indicates a major increase of this element, but such a shift has been found also in samples of brain tissue from a case of Pick's disease and from a recent infarct of the brain of an elderly man. It cannot be considered to be specific for Parkinson's disease, but it suggests a severe alteration of the blood–brain barrier for this element.

We found very little difference in the spectrographic patterns from different topographic areas of the brain, except between gray and white matter as usually shown on wet chemical analyses. In general, if an element was increased or decreased in one area of gray matter, it was increased or decreased in all areas with a few exceptions, such as obvious increase of Ca in areas with calcareous deposits.

No abnormal peaks of Mn, Cu, Pb, or Hg were detected in the Parkinsonian brain.

A vacuum spectrometer can improve the sensitivity of this type of analysis, but the helium tunnel is entirely satisfactory. Helium can prove to be rather expensive since a full tank can be used in a single afternoon when rapid flow is used to replace all of the air for low atomic number elements. Automated systems, multiple sample chambers, and other improvements in instrumentation have greatly simplified and improved X-ray analysis and it seems destined to find wider use in the biological sciences.

4. SUMMARY

X-ray fluorescent spectroscopy was used to determine qualitative and relative amounts of K, S, P, Cl, Ca, Fe, and Zn in specimens of formalin-fixed dried, pelleted brain tissue from 11 Parkinsonian patients and compared with samples of formalin-fixed normal brain and cellulose blanks. The spectrographic profiles revealed an increase of Fe and a decrease of K in the Parkinsonian brain tissue. Gray matter differed from white matter as expected, but there was little difference between various gray or white areas of the brain. The decreased K was shown to be caused wholly or partially by a shift of this element into the formalin during fixation. No abnormal peaks of Mn were detected.

The increased Fe was not specific to the Parkinsonian brain since it was found in formalin-fixed brain tissue in other diseases. This observation requires additional studies on unfixed, uncontaminated, fresh or fresh-frozen brain specimens, but the generalized shift of this element suggests a generalized shift of this element in the Parkinsonian brain tissue. Bromine was detected in one Parkinsonian brain specimen, but no other elements were found in significantly increased amounts. The other elements showed less marked variation, and future quantitative studies with the use of accurate standards or additional methods will be necessary to evaluate these shifts.

X-ray fluorescent spectroscopy has proved to be a valuable instrument for qualitative and semiquantitative determination of all elements from atomic number 14 (Al) through 92 (U) with ordinary commercial equipment (special equipment has been reported to extend the range down to carbon). The limit of detectability in dried, pelletized brain is in the order of 10 $\mu g/g$ for heavy elements and 100 $\mu g/g$ for light elements when ordinary equipment is used, but considerably lower amounts can be determined with the use of special equipment.

REFERENCES

1. K. M. Earle, Studies on Parkinson's Disease Including X-ray Fluorescent Spectroscopy of Formalin-fixed Brain Tissue. *J. Neuropathol. Exper. Neurol.* **27** (No. 1): 1–14 (1968)
2. H. W. Liebhafsky, H. W. Pfeiffer, E. H. Winslow, and P. D. Zemany, In *X-ray Absorption and Emission in Analytical Chemistry*, New York, John Wiley & Sons, Inc. (1960).
3. L. S. Birks, In *X-ray Spectrochemical Analysis*, New York, Interscience Publishers, Inc. (1960).
4. S. Natelson and W. R. Whitford, Determination of elements by X-ray emission spectroscopy; in *Methods of Biochemical Analysis* (D. Glick, ed.) New York, Interscience Publishers, Inc., Vol. 12 (1964) pp. 11–68.
5. H. J. Rose, I. Adler, and F. J. Flanagan, X-ray Fluorescence Analysis of Light Elements in Rocks and Minerals, *Appl. Spectroscopy* **17**, 81 (1963).
6. P. K. Lund and J. C. Mathies, X-ray Spectroscopy in Biology and Medicine. I. Total iron (hemoglobin) content in human whole blood. II. Calcium content of human blood serum. III. Bromide (total bromine) in human blood serum, urine, and tissues. *Norelco Reporter* **7**, 127 (1960).
7. L. Zeitz, X-ray Fluorescence Analysis in Biological Specimens. Symposium on Recent Developments in Research Methods and Instrumentation. (Oral presentation and summary in program.) Clinical Center Auditorium, N.I.H., Bethesda, Maryland, 1966.
8. W. J. Campbell, J. D. Brown, and J. W. Thatcher, X-ray Absorption and Emission, *Anal. Chem.* **38**, 416 (1966).
9. R. C. Hirt, W. R. Doughman, and J. B. Gisclard, Application of X-ray Emission Spectrography to Airborne Dusts in industrial hygiene studies, *Anal. Chem.* **28**, 1649 (1956).
10. S. Natelson, Recent Developments in X-ray Spectrometry as Applied to Clinical Chemistry, *Clin. Chem.* **11** (Suppl.), 290 (1965).
11. J. W. Gofman, O. F. DeLalla, E. L. Kovich, O. Lowe, W. Martin, D. L. Piluso, R. K. Tandy, and F. Upham, Chemical Elements of the Blood of Man, *Arch. Environ. Health (Chicago)* **8**, 105 (1964).
12. D. A. Morningstar, G. Z. Williams, and P. Suutarinen, The Millimolar Extinction Coefficient of Cyanmethemoglobin From Direct Measurements of Hemoglobin Iron by X-ray Emission Spectroscopy, *Amer. J. Clin. Pathol.* **46**, 603 (1966).
13. X-ray Department, General Electric Co., X-ray Wavelengths for Spectrometer, Third edition, Cat. No. A4961, DA, Milwaukee, Wisconsin, General Electric Co. (1964).

14. R. W. Thiers, Contamination in Trace Element Analysis and Its Control, in *Methods of Biochemical Analysis* (D. Glick, ed.), New York, Interscience Publishers, Inc., Vol. 5 (1957), pp. 273–285.
15. L. A. DalCortivo and M. Cefola, Developments in Spectroscopy: Sample Preparation in *Progress in Chemical Toxicology* (A. Stolman, ed.), New York, Academic Press, Vol. 5 (1965), pp. 283–319.

IV. X-RAY EMISSION ANALYSIS IN BIOLOGICAL SPECIMENS

Louis Zeitz

Division of Biophysics, Sloan-Kettering Institute for Cancer Research

X-ray emission techniques which include (a) non-dispersive (b) dispersive–nondispersive and (c) dispersive–dispersive have been used for quantitative assays of certain elements in biological specimens. One or more of these approaches allow for a detectable limit of about 10^{-9} grams of the element of interest for 10 minute integration times for most elements of biological interest from phosphorous (atomic number 15) to iodine (atomic number 53).

The nondispersive and dispersive–nondispersive approaches have been used in assays for zinc content in suspensions of spermatozoa seminal plasma and freeze-microtomed tissue section from the reproductive tract of several mammalian species. Possible applications to other elements and other biological components are discussed. All three approaches can be applied to the determination of the degree of replacement of thymine by bromouracil in analog labeled deoxyribonucleic acid (DNA). The dispersive–dispersive approach is most versatile allowing for replacement determinations with sulfur and chlorine, as well as bromine and iodine, containing analogs.

1. INTRODUCTION

X-ray fluorescence, or more appropriately, emission spectroscopy has been applied in our laboratory in the determination of the concentration of a few select elements in biological specimens. We have used this analytical approach to determine the zinc content of various components of the male reproductive tract of the rat, the dog, and the human. The zinc levels varied from 0.05–3 mg/g dry weight in samples whose masses varied from 20–2000 μg. The form of the specimens was that of a fluid, suspensions, sediment, or tissue. We have also applied it to the determination of the degree of replacement of thymine by bromouracil in analog replaced deoxyribonucleic acid (DNA). These assays have been carried out in connection with studies of enhanced radiosensitivity which accompanies bromouracil replacement. The levels are from 1–50 parts per thousand (ppk) of bromine in samples of 2–20 μg dry mass of DNA. For the sake of this discussion, analysis of samples from 2–2000 μg dry mass will be called "small sample analysis" and is accomplished with either a dispersive–nondispersive ([1]) or a dispersive ([2]) X-ray emission instrument. Dispersive analysis depends on diffraction, or Bragg reflection, from a crystal for wavelength separation or dispersion. Nondispersive analysis refers to energy dispersion by pulse height analysis of the pulse voltages produced by a proportional detector, i.e., a detector having an output pulse voltage amplitude proportional to the energy of the absorbed photon.

A conventional X-ray emission technique ([3]) has also been used in our laboratory for assays of zinc content in human blood serum. This approach is almost identical to that of Gofman ([4]) and will be discussed briefly in Section 5. The zinc in human blood serum is at a level of about 1 ppm wet weight or about 10 ppm of dried serum and the samples are 60 mg of lyophilized serum.

Conventional X-ray emission techniques have been employed by a considerable number of investigators in the determination of element content of various biological material. Gofman ([4]) has carried out an extensive study for many elements in blood plasma. The concentration level of the elements varied from trace elements such as titanium to moderate concentrations for sulfur and chlorine in 400 mg of lyophilized blood plasma. Natelson and co-workers ([5,6]) as well as Lund, Mathies and co-workers ([7]) have applied the techniques of X-ray emission spectroscopy to various problems of interest in the clinical laboratory. Much of the work of these two laboratories involves small sample analysis. The samples are prepared by drying 20–50 μl vol of fluids. Their methods with small samples in some ways parallels our approach. Many others, too numerous to mention, have adopted X-ray emission techniques for the assay of elements found in biological specimens.

2. NONDISPERSIVE AND DISPERSIVE X-RAY ANALYSIS

We have attempted to apply X-ray emission techniques only where the method shows sufficient sensitivity without recourse to chemical separation, separation and collection by ion exchange membranes or ashing. Therefore, the sample preparation procedures are very simple. For small sample analysis, it requires only the proper positioning of specimens of fluids, suspensions, and sediments followed by vacuum drying. Tissue sections are first prepared in the proper shape, freeze-microtomed, positioned in the sample holder, and vacuum desiccated.

Until recently, assays had been carried out using the nondispersive method and the instrument of Hall ([8]). However, this approach was found

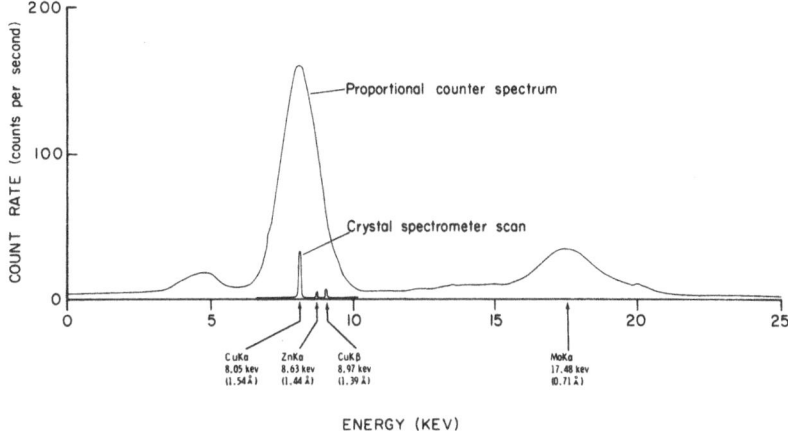

Fig. 1. Comparison of fluorescence spectra taken with a proportional counter spectrometer (nondispersive) and a crystal spectrometer (dispersive). The sample for both spectra is 400 μg sucrose containing 1 μg copper and 0.1 μg zinc.

to have serious limitations in certain types of biological specimens due to interline interferences resulting from the poor resolution inherent in the nondispersive method. The addition of a dispersive spectrometer was the most expedient means of improving resolution while maintaining sufficient sensitivity for assays of elements involved in the studies undertaken. The nondispersive spectrometer was kept for determination of the dry mass. It, of course, could also be used for nondispersive element analysis. At present, we find nondispersive analysis with a proportional counter is of very limited value for element detection since it gives little, if any, advantage in sensitivity for most elements of interest in the type of samples used in our studies (see Section 6). At the same time, it has the disadvantage of a resolution which in some cases is two orders of magnitude inferior to that of the dispersive method. The dispersive and nondispersive spectra of Fig. 1, taken with the same sample of 0.1 μg zinc and 1 μg copper in 400 μg of sucrose, graphically indicates the difference of the two methods in the ability to detect a small amount of an element in the presence of ten times the amount of the adjacent element. A dispersive spectrum of the same type of sample with greater detail in the region of zinc and copper K radiation is shown in Fig. 2.

The arrangement for the dispersive–nondispersive method and a photograph of the specimen chamber, associated crystal spectrometer and proportional counter is shown in Fig. 3. The amount of the element of interest is inferred from the intensity of the element line obtained with the crystal spectrometer. The weight of the sample is determined from the intensity of scattered characteristic Mo K X-ray lines in the exciting

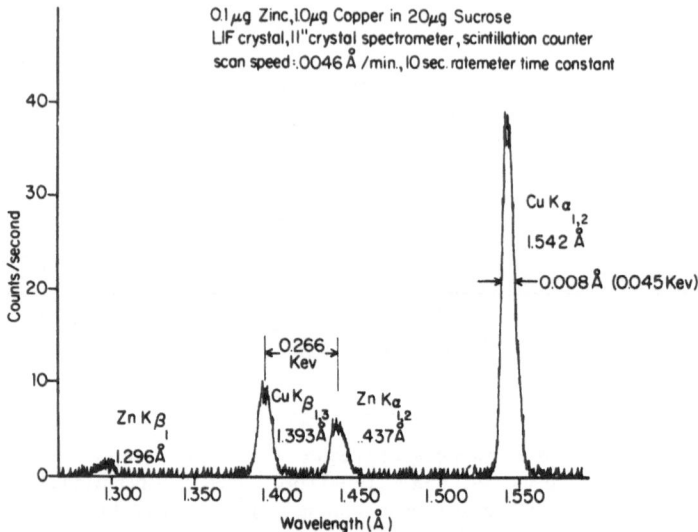

Fig. 2. Spectrum taken with crystal spectrometer of a 20 μg sample of sucrose containing 1.0 μg copper and 0.1 μg zinc.

radiation, since molybdenum target X-ray tubes have been used exclusively for mass determinations. A vacuum specimen chamber is required since the number of Mo K photons scattered from small samples of the order of 30 μg, entering the counter, is about 50 times less than the number entering due to scattering from air at atmospheric pressure ([9]). The vacuum chamber also serves to slightly reduce the background of the crystal spectrometer throughout the spectrum while considerably diminishing it for lines originating in the target or produced as secondary radiation in the specimen chamber. The reflecting type curved crystal X-ray spectrometer uses the Bragg reflection planes bent to a radius of 22 in. and the surface ground to 11 in. according to the method of Johansson ([10]). The motion of the spectrometer maintains focus at all wavelengths available to it and, at the same time, results in a crystal motion which allows the crystal to intercept approximately the same cone of radiation emitted from the sample at all wavelengths ([11]). The detector is either a side-windowed proportional counter or a scintillation counter.

The arrangement and a photograph of the equipment as set up for dispersive analysis is shown in Fig. 4. This arrangement allows for the simultaneous dispersive analysis of elements below atomic number 26 with spectrometer I, and the elements atomic number 26 and greater with spectrometer II. It is also possible to place a proportional counter with its

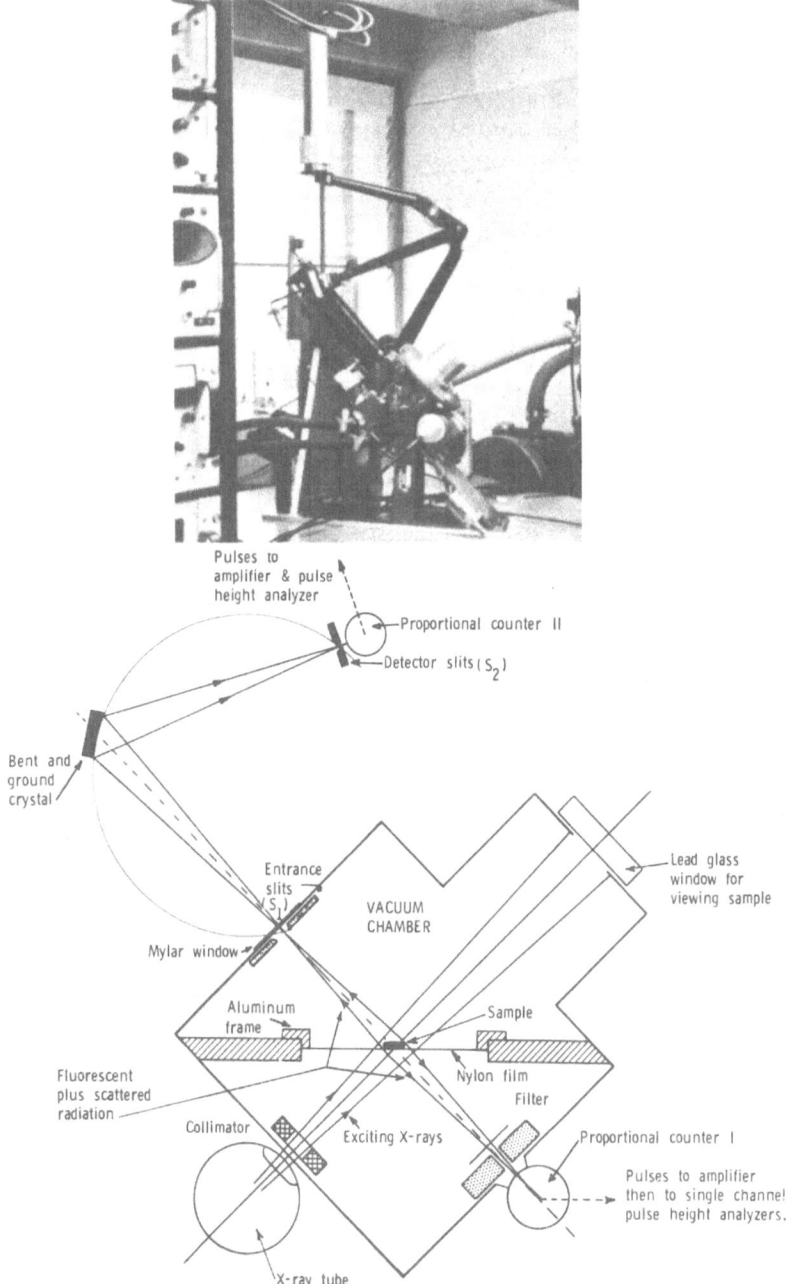

Fig. 3. Arrangement for and photograph of dispersive–nondispersive X-ray emission analysis instrument. Some important distances are: X-ray tube to sample and sample to S, is $2\frac{3}{4}$ in.; sample to collimator of proportional counter I is $\frac{7}{8}$ in.

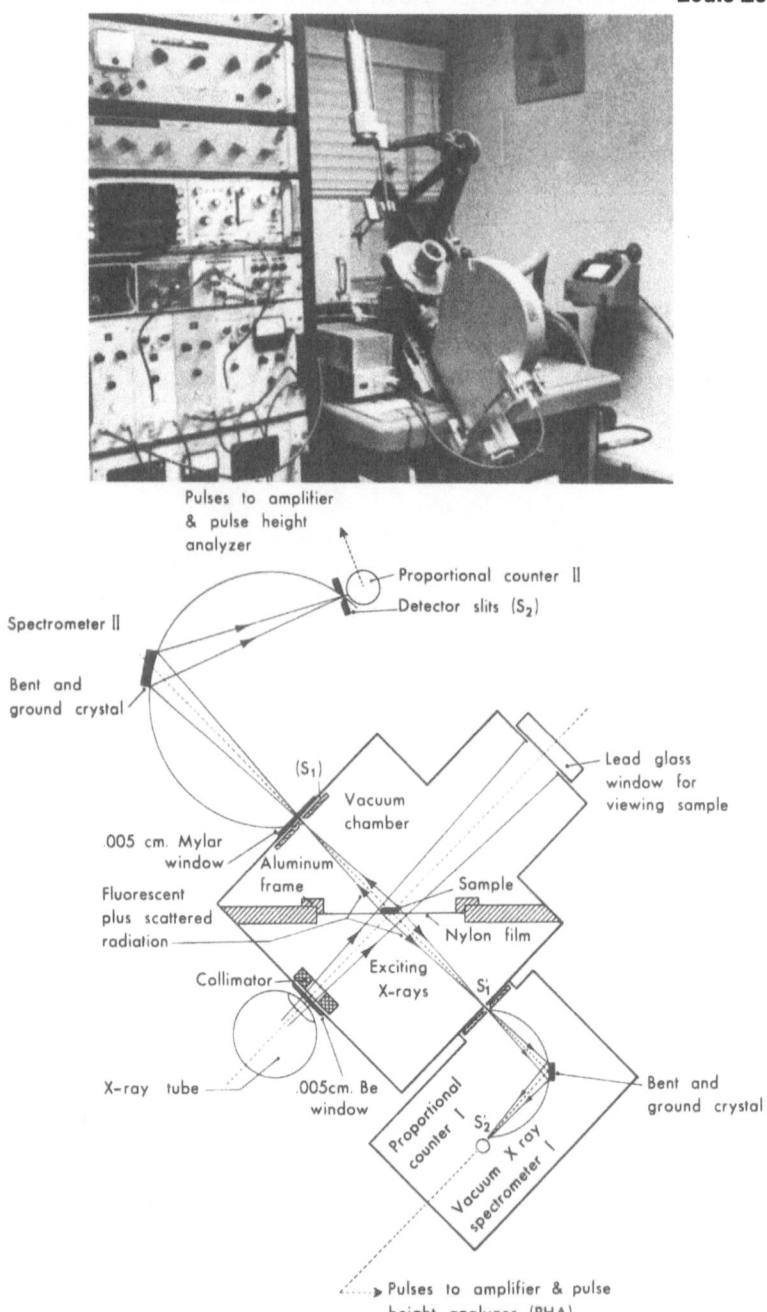

Fig. 4. Arrangement for and photograph of dispersive X-ray emission analysis instrument.

associated collimator and filter holder in the position of slit S_1 of spectrometer II without disturbing the arrangement shown, thus converting it to a dispersive–nondispersive instrument. The curved crystal vacuum X-ray spectrometer (spectrometer I) is essentially the same type as already described for spectrometer II. The Rowland circle radius of this spectrometer is 5.591 in. The detector is either a flow or a sealed-off type proportional counter.

A 50 kV high-voltage supply with milliamp and voltage stabilizer is used to power the X-ray tube. The X-ray tube employed depends on the element under analysis. When maximum sensitivity is required, we use a chromium (Cr) target tube with an 0.25-mm thick beryllium (Be) window in exciting lines whose energy is less than 5 keV, a molybdenum (Mo) target tube with a 1-mm beryllium window for line energies from 5–20 keV and a tungsten (W) target tube with a 1-mm beryllium window above 20 keV.

For small sample analysis, the samples are supported by a single layer of nylon (or polyvinylchloride) from 900–1300 Å thick. The nylon support is prepared by floating a 10% solution of nylon in isobutyl alcohol on water and lifting the film off with an aluminum frame. Polyvinylchloride films are used with acidic samples, of about pH5 or less, since they attack nylon. Figure 5 shows the aluminum frame, sample and frame holder, and positioning template. This method of sample presentation allows us to position the center of the aluminum frame to better than a few thousandths of an inch in the specimen chamber. Samples of solutions and suspension are centered on the nylon film of the aluminum frame by delivering with a micropipet. However, the size and shape of the dried samples cannot be well controlled. The problem of obtaining a reproducible analytical method with this type of sample presentation will be discussed later.

3. DISPERSIVE–NONDISPERSIVE ANALYSIS WITH "SMALL SAMPLES"

It will be noted that the crystal spectrometer in Fig. 3 must detect the fluorescent radiation which is transmitted through the sample. This configuration limits the thickness of samples that can be used. For much of our work, we are dealing with "infinitely thin" samples and using "fluorescence in transmission" is of no concern, as will become evident. With large sample analysis, care must be taken to maintain constant thickness, as well as density and shape, of samples and standards, and to correct for slight thickness changes if they occur.

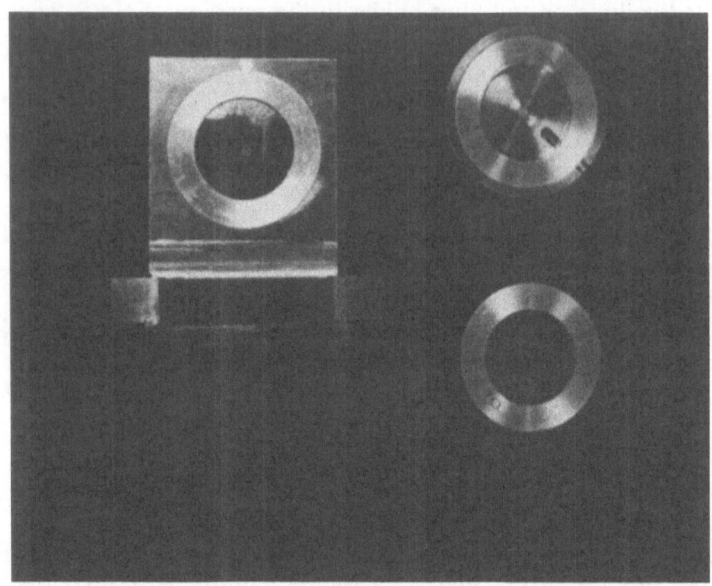

Fig. 5. Aluminum frame to hold nylon film (lower right); frame holder and frame as it goes into specimen chamber with a dried sample in center (left), and aluminum frame with nylon film in centering template (upper right).

To gain insight into the analytical method proposed, the following three aspects must be gone into in some detail:

1. the magnitude of the fluorescence radiation leaving the sample as a function of sample thickness;
2. the validity of using the intensity of scattered lines for the determination of sample mass;
3. the precision and accuracy attainable without reproducing sample size and shape since, in our approach with small samples, the size and shape of samples are not well controlled.

3.1. Sample Thickness and Self-Absorption

An expression can readily be obtained for the magnitude of the fluorescence radiation as a function of sample thickness if one idealizes the geometry and makes several simplifying assumptions. It is felt that although the problem is greatly idealized, the answers so obtained are useful in estimating the errors one can expect due to thickness or absorption effects in the sample.

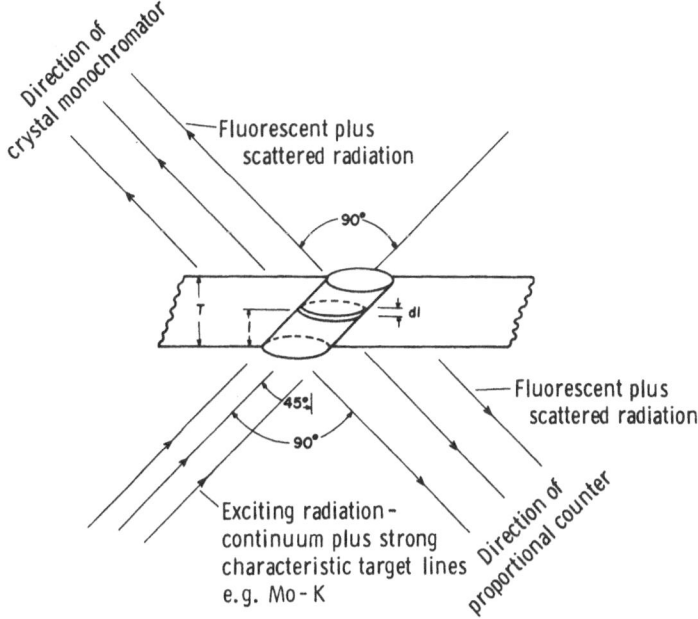

Fig. 6. Geometric configuration of exciting beam of X-radiation, sample and the dispersive and nondispersive detection system.

The geometry of the system is detailed in Fig. 6. We will assume that: (a) the diameter of the exciting pencil of X-rays is very small and non-diverging; (b) the exciting radiation is monochromatic, while in reality use is made of the continuum plus very strong characteristic lines of the element of the target of the X-ray tube; (c) the sample is composed of small concentrations of heavier elements in a light biological matrix; and (d) secondary fluorescence is negligible. Secondary fluorescence, as used here, refers to the excitation of elements in the sample, not by the impinging exciting radiation but by X-rays emitted from other elements within the sample which in turn were excited by the impinging radiation. Assumption (d) follows from the type of sample assumed in (c).

Using these simplifying assumptions and the approach of Compton and Allison ([12]) it can be shown ([11]) that the power of the $K\alpha$ radiation of element A leaving the sample in the direction of the crystal monochromator is given by

$$P(A)_{K\alpha} = \frac{K'M_A}{[(\mu_{\lambda_\kappa}{}^B)_l - (\mu_{\lambda_1}{}^B)_l]} \{\exp[-(\mu_{\lambda_1}{}^B)_l T \sec \theta] - \exp[-(\mu_{\lambda_\kappa}{}^B)_l T \sec \theta]\}$$

where m_A is the mass of element A per unit volume, $(\mu_{\lambda_1}{}^B)_l$ and $(\mu_{\lambda_\kappa}{}^B)_l$

are the linear absorption coefficients, respectively, for the exciting radiation λ_1 and the fluorescent radiation, λ_κ. The constant K' is independent of thickness and

$$K' \propto \frac{P_{\lambda_1} \nu_k (\gamma_k - 1) \Gamma^A_{\lambda_1} \omega_k z_{k\alpha}}{\gamma_k \nu_{\lambda_1}}$$

where only the terms which are a function of the wavelength of the monochromatic exciting radiation λ_1 are of interest in the following discussion. P_{λ_1} and ν_{λ_1} are the power and the frequency of the exciting beam of radiation and $\Gamma_{\lambda_1}{}^A$ is the true atomic absorption coefficient of element A for the wavelength λ_1. The other terms are not of interest in this discussion and their definitions can be found in Refs. 11 and 12. The linear absorption coefficients $(\mu_\lambda{}^B)$ equal $\Sigma(\mu_\lambda^i)_m \cdot m_i$ where $(\mu_\lambda^i)_m$ are the mass absorption coefficients, m_i is the mass per unit volume of element i, and the summation is over all elements except element A, this being neglected because it is assumed present in such a low concentration as to add negligibly to the absorption. This equation, with m_A kept constant as the thickness is varied, is in the form of the difference of two negative exponentials and takes on the shape shown below:

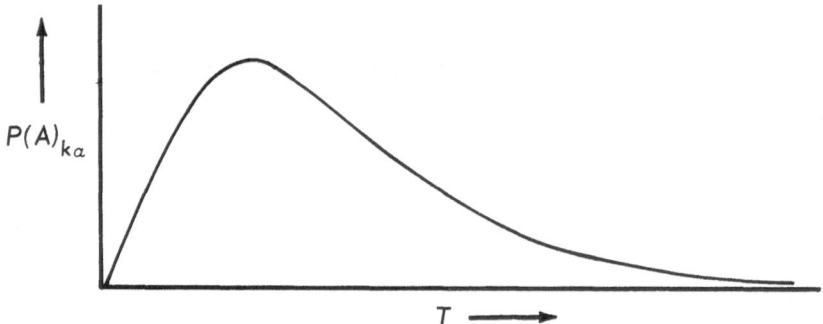

In a qualitative sense this is the shape that is expected in fluorescence in transmission.

In our analytical approach, the size and shape of the sample are not accurately controlled. The thickness varies from sample to sample and within a sample. The necessary condition for a quantitative analytical method with this type of sample presentation is that the magnitude of the detected line must be independent of sample thickness, or stated in another way, that the magnitude of the detected line must be a function of the amount of element present in the sample and not its thickness. We can develop an expression which will indicate the error in accuracy (precision) to expect due to the absorption or thickness effect. This is done by making the substitution

$$m_A = \frac{M_A}{a_b} \cdot \frac{1}{T} = D_A \cdot \frac{1}{T}$$

in the expression already given, where M_A is the total mass of element A in the sample and is kept constant as the thickness is varied, a_b is the area of the base of the cylinder of Fig. 6 and $D_A = M_A/a_b$. This results in the equation

$$P(A)_{K\alpha} = \frac{K'D_A}{[(\mu_{\lambda_K}{}^B)_l - (\mu_{\lambda_1}{}^B)_l]} \cdot \frac{1}{T} \{\exp[-(\mu_{\lambda_1}{}^B)_l T \sec \theta] - \exp[-(\mu_{\lambda_K}{}^B)_l T \sec \theta]\}$$

This expression is used to develop the curves shown in Fig. 7. The ordinate, $P(A)_{K\alpha}/K'D_A$ is essentially the ratio of the power of K radiation of element A actually emitted from the sample into a relatively small, well-defined solid angle to the power which would be emitted into this same solid angle if there were no absorption within the sample. The effective wavelength λ_1 is chosen in a very arbitrary manner. This wavelength is selected so that 10–15% of all the photons in the continuum leaving the X-ray tube (50 kV, 1-mm beryllium window) lie between this wavelength and the K-absorption edge of the element of interest. Obviously, K' will vary from element to element since we are changing the wavelength of the mono-chromatic exciting beam with the element under analysis. The composition of soft tissue is assumed to be that given by Engström [13] which is shown in Table I. The density of dry, soft tissue is assigned a value of 0.5 g/cm³.

All the curves, except the dashed curve for phosphorus in DNA, are based on maintaining the mass of the detected element at 0.1 µg as the thickness is varied. Based on this mass, the maximum fractional weight of detected element in these curves is 0.5% for the 20 µg/cm². The phosphorus concentration decreases with increasing mass per unit area.

It can be shown that similar curves can be constructed for the case of "fluorescence in reflection," or the detection of fluorescent radiation emitted from the same side of the sample as that of the entering exciting radiation (detection with spectrometer I of Fig. 4). Curves for this case vary little from those given in Fig. 7.

A limited amount of experimental data has been obtained which allows for a comparison with calculated results and are given in Fig. 7. There are various technical factors which make it difficult to prepare proper samples for this type of study. The most important involves the inability to obtain the same cross sectional area of dried samples at widely differing dilutions. Only those samples which dry to approximately the

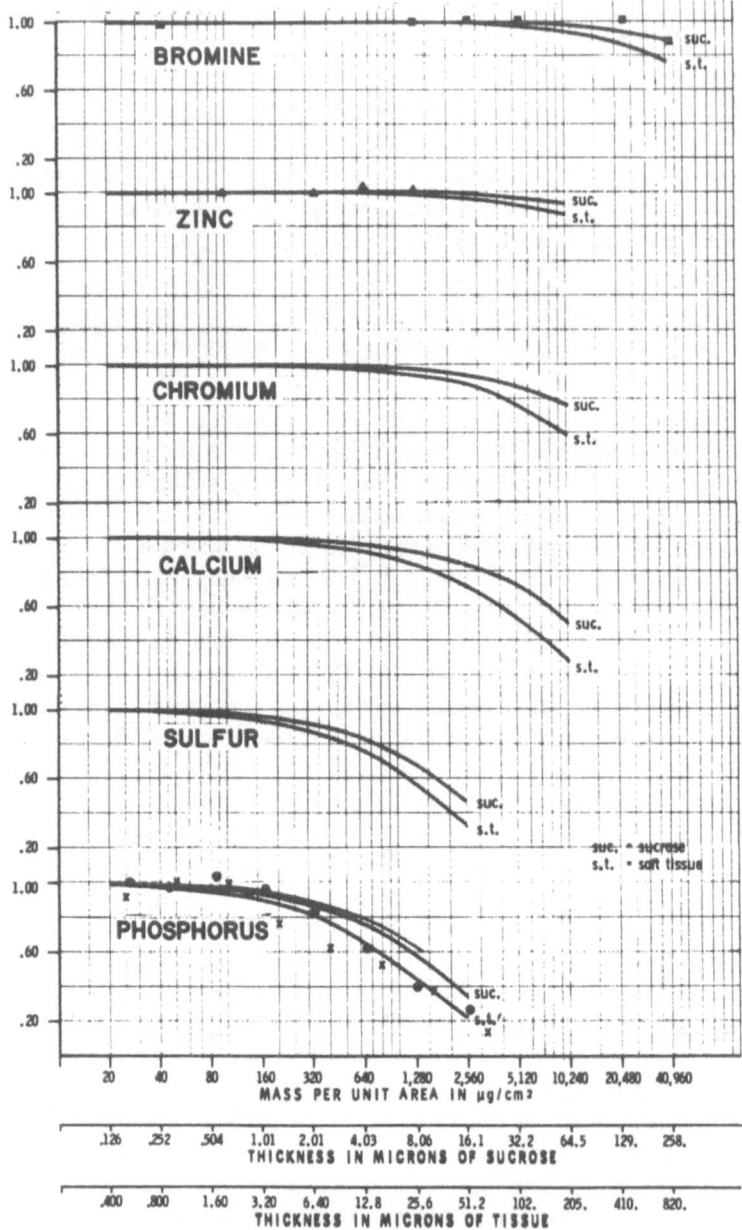

Fig. 7. Self-absorption curves for selected elements. The experimental points are designated x for yttrium ($Y\ L\alpha$) in urine, ○ for phosphorus in salmon sperm DNA and sucrose, △ is for zinc in sucrose and □ for bromine in sucrose.

TABLE I

Proceedings of the 1967 Eastern Analytical Symposium,
L. Zeitz, X-ray Emission Analysis in Biological Specimens

		Fraction by weight			
Element	Sucrose	Soft tissue Lea	Engström	Na salt DNA	Pure DNA
H	0.06	0.04	0.096	0.036	0.042
C	0.42	0.48	0.44	0.353	0.375
N		0.16	0.12	0.157	0.163
O	0.52	0.28	0.20	0.290	0.311
Na		0.004	0.02	0.70	
Mg		0.0016	—		
P		0.008	0.06	0.094	0.102
S		0.008	0.04		
Cl		0.004	0.02		
K		0.014	—		
Ca		0.004	0.004		

same cross sectional area are selected. Even for these samples, there could be variations of thickness within the sample. We are, therefore, dealing with a crude average weight per unit area. In our judgment the limited experimental data obtained are in sufficient agreement to justify use of the calculated curves as a guide in determining the maximum sample thickness for use in quantitative work. It is evident from the self-absorption curves that thinner samples must be used as we go to the lighter elements, with sample thickness becoming a very limiting factor in analyses for elements whose atomic number is that of sulphur or below.

3.2. Measure of Mass from Magnitude of Scattered Line

The sample mass is determined from the magnitude of scattered lines detected by proportional counter I in Fig. 3. This is justified only if the scattering per unit mass is nearly constant for biological samples. If an organic molecule such as sucrose is used as a standard for determining mass, it must differ little in its scattering per unit mass from biological material. The scattering per gram from a noncrystalline sample is

$$I_{sc} = \sum_{i=1}^{n} I_a \frac{N}{A_i} C_i = N \sum_{i=1}^{n} I_a \frac{C_i}{A_i}$$

where I_a is the scattered intensity per atom, C_i is the weight fraction of element i, A_i is the atomic weight of element i, N is Avogadro's number and the summation is for all the n elements present; and

$$I_a = I_e \left\{ \left(\sum_1^z f_n \right) + \left[z - \left(\sum_1^z f_n^2 \right) \right] \right\}$$

where I_e is the intensity scattered by a free electron, z is the atomic number of the element and the two summations of the structure factor of the nth electron are available in the literature, tabulated for various atoms ([12]).

Using the elemental compositions of Table I, the scattering to expect from various samples used in our work is shown in Table II. The scattered lines, Cr K, Mo K and Ag K, are chosen because these are some of the targets of X-ray tubes generally used in fluorescence analysis. We have used only molybdenum target tubes in determining the mass of sample from the scattered target lines. The values given for Ag K are based on extrapolated values for summation of the structure factors since the tables do not give these values for sin $(\phi/2)/\lambda$ as large as 1.27. The large error 8% that is expected with a sodium salt of DNA containing 5% bromine is of little concern in our application for two reasons. First, a bromine weight fraction of 5% is equivalent to a replacement of thymine by bromouracil of about 100% ([9]). In these replacement studies, we have never been concerned with replacement of greater than 50%. Secondly, in this assay it is known that bromine is the only heavy element whose concentration is changing to any appreciable degree, and it is the increase in bromine concentration which is contributing to the inordinately high scattering for this sample. Therefore, we can make a mass correction based on the intensity of the Br $K\alpha$ line detected with the crystal spectrometer.

A sample with an elemental composition similar to that given by Engström for soft tissue should result in an error in mass determination of about 7%, if sucrose is used as a standard. If this error is felt to be too large, one could homogenate the tissue under study and use a carefully weighed sample of this homogenate as a standard. The presence of an appreciable amount of crystalline material in the tissue would introduce an error in a value for the mass determined on the basis of X-ray scattering. Therefore, this approach can not be used with tissues, or any other samples, having an appreciable degree of crystallinity.

The calibration curve of Fig. 8, used to determine the mass of a sample, is obtained with sucrose or DNA standards. Straight line mass calibration curves, similar to that of Fig. 8, have also been obtained with dried urine standards. The magnitude of the scattered molybdenum line is N'_{Mo}, where $N'_{Mo} = N_{Mo}^s$ (counts due to variously scattered Mo-K photons from sample, nylon film and specimen chamber) $-N_{Mo}^B$ (counts due to scattered Mo-K photons from nylon film and chamber). With small samples, where

TABLE II

Proceedings of the 1967 Eastern Analytical Symposium, L. Zeitz, X-ray Emission Analysis in Biological Specimens

Sample	Scattered intensity (relative)			Ratio of scattered intensities (sucrose scattering normalized to 1.00)		
	Cr-K (2.29 Å) $\frac{\sin(\phi/2)}{\lambda} = 0.31$	Mo-K (0.71 Å) $\frac{\sin(\phi/2)}{\lambda} = 1.00$	Ag-K (0.56 Å) $\frac{\sin(\phi/2)}{\lambda} = 1.27$	Cr-K	Mo-K	Ag-K
Sucrose	1.035	0.581	0.568	1.00	1.00	1.00
Soft tissue { Engström	1.151	0.620	0.583	1.11	1.07	1.03
Lea	1.025	0.579	0.554	0.99	1.00	0.98
Lea + 0.5% Zn		0.584			1.01	
DNA	1.084	0.564	0.533	1.05	0.97	0.94
Na-salt of DNA	1.192	0.576	0.543	1.15	0.99	0.96
Na-salt of DNA + 5% Br		0.626			1.08	

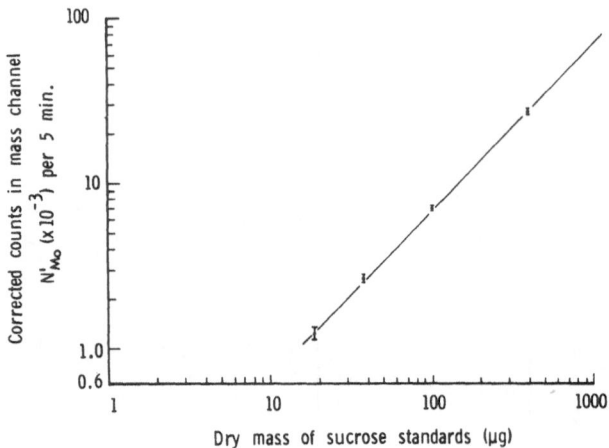

Fig. 8. Calibration curve for mass using "precalibrated" nylon films and 10 μl vol of sucrose solution. The indicated points and standard deviations for the 18 and 38 μg masses are the mean and standard deviations based on measurements of 6 separate 10 μl samples in each case. The other standard deviations are estimated from these.

the intensity of the signal is not much larger than the background intensity, small errors in the determination of the background intensity introduce large errors in the mass determination. This error can be diminished with small samples by "precalibration" of the nylon films for each sample. Precalibration is accomplished by measuring N_{Mo}^{B} for each film before placing the sample on it.

3.3. Reproducibility

Can we expect to obtain an adequate degree of precision and accuracy without reproducing the size and shape of samples accurately? We have attempted to obtain an answer experimentally. The experimental results, as given in Fig. 9, suggest that the sample dimensions must be kept below an experimentally determinable maximum size if the magnitude of the element line is to be a function only of the amount of element present and not of the sample size. The same is true for the magnitude of the line in proportional counter I (Fig. 3) for the determination of mass. This data suggests that size variations of samples below ¼ in. will introduce an error of about 4% in the magnitude of the Zn $K\alpha$ line. There is a compensatory effect in taking the ratio of N'_{Zn}/N'_{Mo} and the error here should only be about 2% for samples ¼ in. or less.

The precision actually obtained by analyzing six separate samples of 10 μl of seminal plasma of dog for 10 min. integrations gave approximately

4, 3, and 2%, respectively, for the relative standard deviations of N'_{Zn}, N'_{Mo}, and N'_{Zn}/N'_{Mo} where the relative counting error expected on the basis of counting statistics is 1.2, 0.8, and 1.4%.

A log–log plot of the ratio of corrected counts in the element channel to corrected counts in the mass channel versus actual mass of zinc to mass of

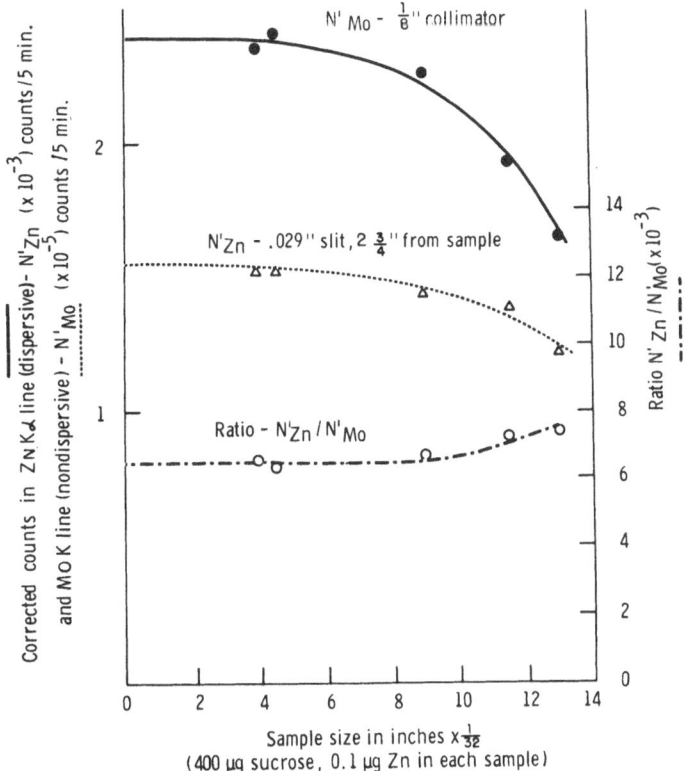

Fig. 9. Corrected counts in zinc Channel N'_{Zn}; corrected counts in molybdenum channel N/N'_{Mo}, and their ratio N'_{Zn}/N'_{Mo} as a function of sample size. N'_{Zn} is obtained from the relation:

$$N'_{Zn} = N_{Zn}{}^S - \left(N_{Zn}{}^B + \frac{\Delta N_{Zn}{}^{Bt}}{\Delta M} \cdot M_s\right)$$

where $N_{Zn}{}^S$ is the measured intensity of the Zn $K\alpha$ line for a 400-μg sucrose sample containing 0.1 μg zinc, $N_{Zn}{}^B$ is measured intensity at the position of the Zn $K\alpha$ line for the nylon film, $\Delta N_{Zn}{}^{Bt}/\Delta M$ is the rate of change of Zn $K\alpha$ counts with mass of pure sucrose standards, and M_s the mass of the sample by an X-ray determination. N'_{Mo} is the difference in counts in the molybdenum channel with 400-μg sucrose sample and with no sample, nylon film only.

sample in accurately prepared standards is given in Fig. 10. The fact that a straight line approximates the data points so accurately indicates that the dispersive–nondispersive method developed can be used with "small samples" to determine the ratio of the mass of the element of interest to that of the sample for biological specimens, i.e., specimens with a small percentage of a heavy element in a light biological matrix.

3.4. Resolution-Sensitivity, Precision, and Accuracy

An idea of the resolution at maximum sensitivity of the air spectrometer (spectrometer II in Fig. 4) can be inferred from Fig. 2. The spectrum shown was taken with the slits adjusted for optimum sensitivity by adjusting

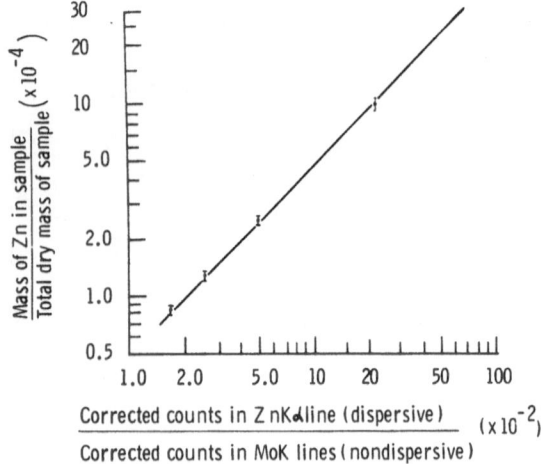

Fig. 10. Ratio N'_{Zn}/N'_{Mo} versus the ratio of mass of zinc to mass of sample in accurately prepared standards of zinc in sucrose. N'_{Zn} and N'_{Mo} are defined in the legend of Fig. 7. All the standards contain 0.1 μg of zinc in varying masses of sucrose and all samples are placed down in 35 μl vol. The points and standard deviations are (1) determined on the basis of measurements of 6 separate 0.1 μg zinc in 400 μg sucrose standards for the 2.5×10^{-4} ratio, (2) based on measurements of 2 separate 0.1 μg zinc in 100 μg sucrose for the 10^{-3} ratio, (3) based on single measurements of 0.1 μg zinc in 800 and 1200 μg sucrose for the 1.25 and 0.83×10^{-4} ratio samples, respectively, where the relative standard deviation is assumed equal to that found in the case of 0.1 μg zinc in 400 μg sucrose standards, and (4) based on a single measurement on the 0.1 μg zinc in 40 μg sucrose standard for the 25×10^{-4} ratio, with no estimate of the level of confidence given.

the slit widths for the maximum line squared over background, L^2/B. This occurs for the Cu $K\alpha$ line with $S_1 = S_2 = 0.029$ in. and the full width at half maximum for this line with these slit widths is 0.008 Å, or in terms of percent of the wavelength or energy of the line it is 0.5 %. This resolution allows for the detection, without interference, of 0.1 μg of zinc in the presence of ten times this amount of the adjacent element copper. This degree of resolution is sufficient for all the analyses we have undertaken. If interline interferences are a problem, the slit widths could be decreased to gain resolution at the expense of sensitivity.

The spectra of Fig. 11 are taken with the vacuum spectrometer operated at close to maximum sensitivity (S_1' in Fig. 4 is 0.015 in. with no slit at S_2'). They indicate that interline interferences are not a problem in biological material for the adjacent elements sulfur, chlorine, and phosphorus even when all three are present in considerable amounts (Fig. 11b).

An estimate of the limits of detectability can be made with the aid of the calibration curves given in Fig. 12. Loss of detectability is arbitrarily assigned to a line whose magnitude is less than four times the magnitude of the counting error in the background. The limit of detectability for various elements determined on the basis of the calibration curves of Fig. 12 is tabulated in Table III. A limit of detectability in grams of element present for a 10-min. integration is given for small samples ($\sim$ 10 μg) as well as for a mass of sample which is considered to be a maximum mass sample for negligible self-absorption, chosen at about 2 %, as determined from the curves of Fig. 7. Also included are estimates of detectable limits in parts per million dry weight for bromine and zinc in large sample analysis (see Section 5) where the sample is 60 mg of lyophilized blood serum.

The results of precision analyses on the dispersive–nondispersive system are summarized in Table IV. Precision studies for the determination of bromine to phosphorus ratios, required in determining the degree of replacement in DNA (see Section 4.3.), are also included. The relative standard deviations in each case are based on the counts obtained for six separate samples prepared from the same solution. The values of the relative counting errors are those expected on the basis of counting statistics where they take into account the error added due to subtraction of the background from the signal. Evidence obtained from repositioning studies with the same sample indicates that the major portion of the increase in actual error over the counting error, which is evident in this precision study, results from the inability to exactly reproduce sample position.

Only a very preliminary evaluation of the accuracy of our analytical techniques has been carried out. A comparison of zinc determination by dispersive–nondispersive small sample analysis and atomic absorption was

TABLE III

Proceedings of the 1967 Eastern Analytical Symposium, L. Zeitz, X-ray Emission Analysis in Biological Specimens

	Max. mass negligible self-abs. of line in sucrose μg	Counts from element c/μg/10 min (c/PPM/10 min)	Background Mass c/μg/10 min	Background Nylon film c/10 min	Total max. Mass sample c/10 min	Limits of detectability PPM dry wt. Max. mass sample 10 min integration	Limits of detectability Grams of element in small samples ($\sim$ 10 μg)
Br Kα 1.041 A (Br-Kα)	4,500 (60 mg dry serum)	1.3×10^5 (3.5×10^3) c/PPM/10 min	5.7	180	2.6×10^4 (1.2×10^5)	1.1 (0.4)	4×10^{-10}
ZnKα 1.437 A (Zn-Kα)	2,100 (60 mg dry serum)	4.7×10^4 (1.0×10^3) c/PPM/10 min	1.4	50	3.0×10^3 (2.4×10^4)	2.2 (0.6)	7×10^{-10}
ILα 3.148 A	240	2.5×10^4	0.3	50	1.2×10^2	7.0	1×10^{-9}
CaKα 3.360 A	240	4.7×10^4	0.5	82	2.0×10^2	5.0	8×10^{-10}
SKα 5.373 A	60	6.1×10^3	0.3	30	48	77	4×10^{-9}
PKα 6.155 A	45	3.2×10^3	—	30	30	152	7×10^{-9}

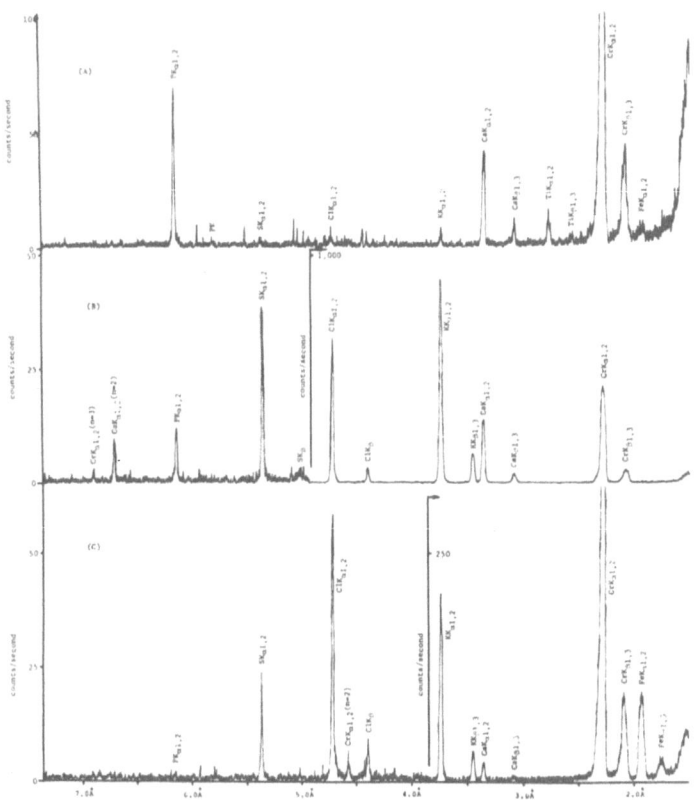

Fig. 11. X-ray dispersive emission spectra of (a) about 150 μg of salmon sperm DNA; (b) 10 μl sample of urine; and (c) 20 μl sample of blood. Samples are prepared by placing the specimen on about 1000 Å nylon film and drying with an infrared lamp for 10–20 min. Entrance slit: S'_1 of Fig. 3 is 0.015 in. with no slit at S'_2. The % FWHM for Ca, K, Cl, S and P $K\alpha$ lines are 0.81, 0.67, 0.44, 0.34, and 0.30%, respectively.

made on dog seminal plasma specimens. In this limited comparison, the two methods gave values for zinc content which agreed to better than 5% ([1]). Also, values of zinc content of human blood serum obtained by our X-ray emission methods with large samples (Section 5) were compared to those obtained by neutron activation analysis. The agreement here was better than 3% ([14]).

TABLE IV

Proceedings of the 1967 Eastern Analytical Symposium, L. Zeitz, X-ray Emission Analysis in Biological Specimens

	Relative standard deviations	Relative standard counting errors
Six 20 μl samples −0.1 μg Zinc/400 μg sucrose 5 min integration	$N'_{Zn}-5\%$, $N'_{Mo}-4\%$, $N'_{Zn}/N'_{Mo}=6\%$	$N'_{Zn}-3\%$, $N'_{Mo}-0.2\%$, $N'_{Zn}/N'_{Mo}-3\%$
Six 10 μl samples of dog seminal plasma 5 min integration	$N'_{Zn}-4\%$, $N'_{Mo}-3\%$, $N'_{Zn}/N'_{Mo}=2\%$	$N'_{Zn}-1.2\%$, $N'_{Mo}-0.8\%$, $N'_{Zn}/N'_{Mo}-1.4\%$
Six 10 μl samples −0.8% dry weight Br } 10 μg DNA 8% dry weight P 20 min. integration	$N'_{Br}/N'_{P}-3.7\%$	$N'_{Br}/N'_{P}-2.5\%$
Six 5 μl samples −1.7% dry weight Br } 2 μg DNA 8% dry weight P 20 min integration	$N'_{Br}/N'_{P}-5.9\%$	$N'_{Br}/N'_{P}-3.4\%$

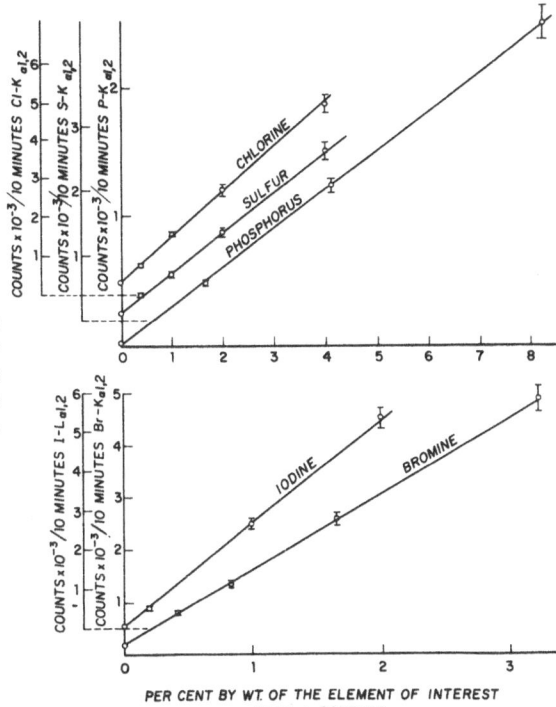

Fig. 12. Calibration curves for Phosphorus (P), sulfur (S), chlorine (Cl), iodine (I, I $L\alpha$ line) and bromine (Br) in DNA or DNA simulated matrix.

4. TYPICAL RESULTS IN BIOLOGICAL MATERIAL

4.1. Zinc Content of Sperm Suspension

A recovery study was carried out ([15]) on a canine ejaculated sperm suspension. The seminal plasma from 6.0 ml of semen was separated by centrifugation. The sperm were then washed with Krebs–Ringer buffer solution and again taken up in Ringers. Two equivalent sperm suspension samples were prepared by the addition of 0.4 ml of zinc-free physiological saline solution to each of two 1.6 ml aliquots of sperm suspension. In this study, as well as in all the studies of zinc in sperm, we use Ringer's solution, physiological saline and distilled water in which the zinc level is ascertained to be less than our limit of detectability, and all glassware is treated to maintain zinc-free conditions. The total zinc content of each of these two equivalent suspensions was calculated by determining the absolute amount of zinc in accurately measured aliquots from each of them. The suspension remaining after removal of 20 μl were designated A and B and were treated as diagramed:

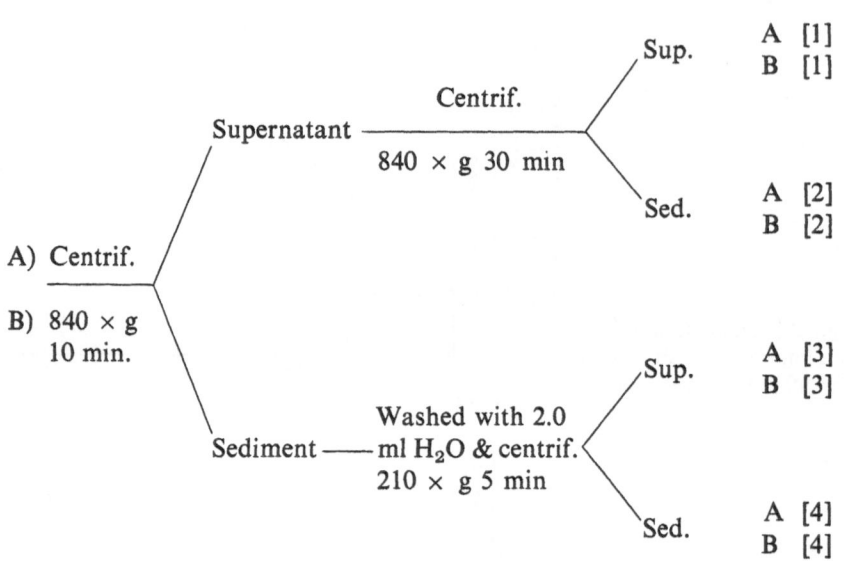

TABLE V

Sample	μg Zn ± estimated error, μg[a]	Weight fraction, mg Zn/g dry wt.
A	3.1 ± 0.3	
A [1]	1.2 ± .18	5.0
A [2]	0.2 ± .03	0.7
A [3]	0.8 ± .32	30.
A [4] Part I[b]	0.7 ± .08	0.9
A [4] Part II	0.6 ± .06	0.8
A [4] Part III	—	—
Sum of zinc content in components	3.5 ± 0.4	
B	2.7 ± 0.3	
B [1]	1.0 ± .16	4.0
B [2]	0.2 ± .03	0.7
B [3]	0.5 ± .30	22.
B [4] Part I	0.7 ± .08	0.8
B [4] Part II	0.7 ± .08	0.8
B [4] Part III	—	—
Sum of zinc content in components	3.1 ± 0.4	

[a]The errors are given in terms of the ±95% confidence limits.
[b]The sediment remaining in the bottom of the centrifuge tube was taken up in a small volume of water (about 25 μl) to make the sample (A[4] part I) for a zinc determination. The centrifuge tube was then washed with successive 25 μl volumes of water until there was no detectable zinc in the washing water. These successive washings are used as samples and designated A [4] part II and A [4] part III. In the duplicate runs, it required a third washing in both cases (A [4] part III and B [4] part III) before the zinc was no longer detectable.

The zinc content of each of the four components of suspensions A and B was calculated from determinations of the zinc content in accurately measured volumes in the case of the supernatant solutions, or from analysis for the weight of zinc in the total sediment. The results given in Table V suggest that, within the estimated experimental errors, the zinc content of the sum of the components equals the content in the original suspension for both specimens analyzed. Also, the zinc determinations for the two equivalent suspensions, and for the respective components of those suspensions, suggests agreement to within the estimated experimental error. If we assume the given confidence limits represent random errors, we can state the following: (a) differences in determinations of the sum of A components and A, sum of B components and B, A[1] and B[1], A[2] and B[2], A[3] and B[3], and A[4] part I and B[4] part I are not significant at about the 5% level: and (b) differences in determinations of A and B, and A[4] part II and B[4] part II are not significant at about the 2% level.

It is of interest to note that the weight fraction of zinc in the sediment components [2] and [4] vary between 0.7 and 0.9 mg zinc/g dry weight. This is the range of weight fractions found in other canine ejaculated sperm which are prepared for assay by the same procedure as that used to obtain component [4]. Sediment [4] is primarily ejaculated sperm. The zinc weight fraction in sediment [2] suggests that it is also primarily whole or fragments of ejaculated sperm. If this is the case, then additional washing in zinc-free solvent (parenteral water), as was done in the case of component [4] as compared to [2], has not affected the fraction by weight of zinc in dried canine ejaculated sperm.

Studies of the zinc content of spermatozoa from various levels of the canine and rat reproductive tract ([16]) have been carried out. The samples for these analyses were treated in a manner similar to that described in the recovery experiment. The zinc assay is carried out after repeated washing with Ringer's solution and water. The final form of the specimen is that of a small sediment at the bottom of a centrifuge tube. An indeterminate volume of the sediment is taken up with a Pasteur pipette and dropped on a thin nylon film to act as a sample. All measurements of zinc content are determined as a ratio of zinc mass to dry mass of sample as determined by a measure of the scattered Mo K lines. The dry masses of these samples, as determined from the X-ray measurement, varied from 40–2000 μg. The zinc content found in canine "preprostatic" sperm is given in Table VI.

The zinc content of ejaculate sperm of dog was also determined ([16]). The ejaculate sperm was obtained manually and, thereafter, treated identically to the preprostatic sperm. The zinc content of ejaculated spermatozoa from six different dogs are shown in Table VII, together with the

TABLE VI

**Proceedings of the 1967 Eastern Analytical Symposium,
L. Zeitz, X-ray Emission Analysis in Biological Specimens**

	Zinc content of spermatozoa mg Zn/g dry weight		
Dog	Testis	Epididymis	Vas
233	0.11	0.10	0.12
549	0.10	0.17	0.11
298	0.11	0.23	—
550	0.22	0.15	0.29
478	0.25	0.14	0.19
580	0.13	0.15	0.17
Mean	0.15	0.16	0.18

TABLE VII

**Proceedings of the 1967 Eastern Analytical Symposium,
L. Zeitz, X-ray Emission Analysis in Biological Specimens**

	Spermatozoa			Seminal plasma		
Dog	Number of specimens	Mean ± SD*	Range	Number of specimens	Mean ± SD*	Range
344	5	1.95 ± 0.45	1.47–2.41	5	2.05 ± 0.22	1.78–2.38
754	5	0.69 ± 0.12	0.53–0.80	5	1.77 ± 0.12	1.62–1.90
694	5	0.60 ± 0.09	0.47–0.68	5	1.47 ± 0.56	0.81–2.13
273	2	0.54	0.52–0.57	2	1.70	1.60–1.80
559	2	0.64	0.54–0.73	—	—	—
205	1	1.70	—	1	1.73	—
Total	20	1.01	0.47–2.41	18	1.75	0.81–2.38

*Standard deviations are given where it is meaningful.

zinc contents of the corresponding seminal plasmas. The mean value of ejaculate is 6–7 times higher than the means of preprostatic sperm. The data indicates that most of the zinc in ejaculate canine spermatozoa is derived from prostatic fluid. The data obtained also showed a considerable variation in the zinc content of ejaculated spermatozoa, whereas zinc concentrations of preprostatic spermatozoa was relatively constant.

4.2. Zinc Content of Tissue Sections

A study (unpublished) in collaboration with the Urology Service of Memorial Hospital was undertaken to determine the effect of different chelating agents on the zinc level of rat testis. The testis was prepared for zinc assay by coring the proper size cylinder, cutting 50-μ thick sections with a freeze-microtome, positioning the samples on a thin nylon film and vacuum dessicating.

Again, all determinations are in terms of the ratio of zinc mass to dry mass of the sample. The zinc concentration found in the interior of the rat testis treated with various chelating agents, as well as that of the controls is given in Table VIII. The two data in the same row are for adjacent 50-μ testis sections and going down the columns, sections from different animals. Considering the variation that may exist, within one rat testis and from rat to rat, the reproducibility of analyses of adjacent tissues is good. The instrumental precision for the detection of zinc in tissue sections cannot be determined from this data. However, the data indicate that the instrumental precision, as measured by percent standard deviation, is better than about $\pm 20\%$. This follows from a calculated percent standard deviation of about 20% for all the EDTA data treated as a single population coupled

TABLE VIII

Proceedings of the 1967 Eastern Analytical Symposium,
L. Zeitz, X-ray Emission Analysis in Biological Specimens

	Concentration of zinc in parts/1000 duplicate analyses on adjacent tissue		Mean of duplicate determinations	Overall average
Dithizone treated—	0.05	0.13	0.09	
50 mg/kg in one injection.	—	0.09	0.09	0.07
Animal sacrificed 3 days after injection	0.06	0.08	0.07	
EDTA treated—	0.15	0.20	0.18	
2000 mg/kg sacrificed	0.09	0.18	0.14	
after 3 days	(0.21)	(0.18)	0.20	0.17
	0.20	0.17	0.19	
	0.13	0.14	0.14	
DDC treated—				
1250 mg/kg sacrificed	0.11	0.11	0.11	0.11
after 3 days				
Controls	0.13	0.12	0.13	
	0.16	0.21	0.19	
	0.13	0.10	0.12	0.15
	(0.18)	(0.18)	0.18	
	0.13	0.17	0.15	

with the fact that the variance in the data due to instrumental error alone must be less than the overall variance, which includes intrinsic variation of zinc content in the animals.

The mean of the zinc content for the dithizone treated rats is significantly different from the means of the controls and those treated with EDTA. Of the 5 DDC treated rats, one animal survived for three days after treatment. In the case of the EDTA treated rats, 3 survived. This explains the limited data shown for these two types of treatment.

4.3. Bromouracil Replacement Determination

A number of laboratories have reported finding changes in radio-sensitivity in bacteria, virus and tissue cell cultures resulting from replacing thymine by bromouracil in the DNA of the organism. Studies of this effect are being carried out in the Biophysics Division of Sloan-Kettering Institute. In a study of this nature, it is essential to establish the degree of replacement of thymine by its analog bromouracil (BU). A nondispersive X-ray emission technique has been used to determine the degree of replacement. This approach is already described in detail ([9]) in the literature and will not be further discussed here. Our present method allows for the determination of the degree of replacement by an assay for the ratio of bromine to phosphorus atoms present in the sample. The accuracy of the analysis with this approach becomes a function of the amount of phosphorus present in non-DNA material which can be considered as impurities in the DNA preparation. The bromine line is detected with the air spectrometer (spectrometer II of Fig. 4) and the phosphorus line either nondispersively with the proportional counter as depicted in Fig. 3 or dispersively with the vacuum spectrometer (spectrometer I in Fig. 4).

The degree of replacement can be obtained by an assay for the ratio of the number of bromine (Br) to phosphorus (P) atoms present, since each nucleotide of DNA contains one phosphorus atom. It can be shown that the fractional replacement F of thymine by bromouracil is given by the expression:

$$F = \frac{f_{\text{Br}}}{f_T} = \left(\frac{1}{f_T}\right)\left(\frac{1}{C}\right)\left(\frac{N'_{\text{Br}}}{N'_{\text{P}}}\right)$$

where f_{Br} is the fraction of the bases in the labeled DNA which are BU, f_T is the fraction of bases in the unlabeled DNA which is thymine, $C = K_{\text{Br}}/K_{\text{P}}$ where the K_{Br} and K_{P} are the counts per atom of bromine and phosphorus, respectively, and the corrected counts, N'_{Br} and N'_{P}, are the counts due to bromine and phosphorus, respectively, present in the sample. The values for the slopes of the calibration curves, K_{Br} and K_{P}, are obtained

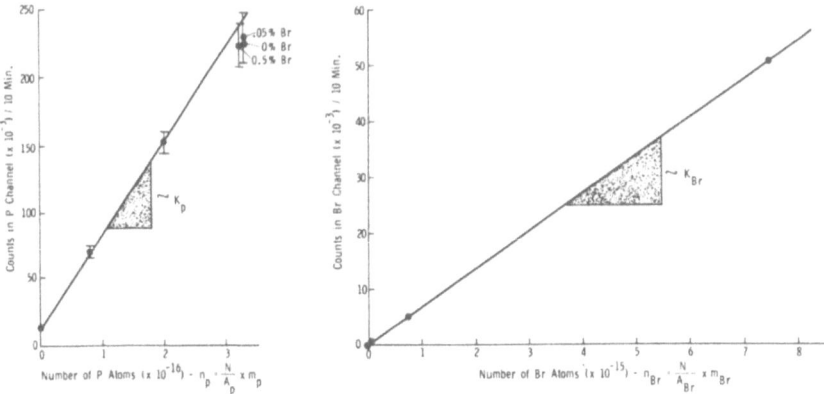

Fig. 13. Calibration curves for phosphorus (P) and bromine (Br) in DNA. The phosphorus calibration curve developed from standards prepared with salmon sperm DNA, bromodeoxyuridine (BUdR) and thymidine. All samples prepared by delivering 20 μl of a solution containing 1 mg dry mass/ml of solution. Phosphorus is detected nondispersively with a proportional counter having a 0.002-in. beryllium window and bromine dispersively with the air crystal spectrometer employing a LiF crystal and a scintillation counter as the detector. A Machlett AEG-50-S molybdenum target X-ray tube operated at 50 kV, 30 mA was employed.

from working curves such as those shown in Fig. 13. The abscissa of the calibration curves represents the number of atoms of a particular element present in the sample. It is obtained by multiplying the X-ray determination of the mass of the element by the number of atoms per unit mass, N/A, where N is Avogadro's number and A is the atomic weight of the element. The corrected counts are obtained from the difference $N'_{Br} = N_{Br}{}^{s} - N_{Br}{}^{B}$, where $N_{Br}{}^{s}$ is the count in the Br $K\alpha$ line for a 20-μg sample and $N_{Br}{}^{B}$ is the measured count in the position of the Br $K\alpha$ line for the nylon film plus a 20-μg sample of DNA containing no bromine. The same approach is used for N'_{P}, but in this case 20-μg of thymidine is used as a sample to obtain a background count. All calibration curves are obtained with drift corrected data. This correction is generally very small and at times negligible. The same approach can be applied to replacement determination other than BU.

The error of the replacement determination with this technique, excluding the ordinary instrumental and counting statistical errors, varies directly with the mass of phosphorus present in the non-DNA material of the sample, e.g., ribonucleic acid (RNA) or phospholipids, but within limits this error does not depend on the mass of non-phosphorus containing impurities.

The number of phosphorus or bromine atoms in the sample can be determined from a measure of the magnitude of the appropriate spectral line intensity obtained either dispersively or nondispersively. The non-dispersive approach cannot be used with samples of isolated DNA which give interline interferences in the vicinity of the analytical line of interest. Use of this method with such samples requires recourse to complicated spectral stripping techniques ([17]). The spectrum of salmon sperm DNA (obtained from California Biochemicals Corp., the same material analyzed in Fig. 11a) given in Fig. 13 suggests that for this sample the magnitude of the phosphorus-K lines can be determined by nondispersive analysis with an adequate degree of accuracy for many studies involving replacement. Frequently, isolated DNA contains elemental impurities which introduce serious interline interferences. This is the case for the HeLa cell DNA* of Fig. 14 making it impossible to obtain an accurate measure of the magnitude of the phosphorus-K lines in a straightforward manner by means of non-dispersive analysis. A dispersive approach can be applied to this type of sample as is evident from spectrum A in Fig. 11.

A ratio of the counts in the bromine to that in the phosphorus channel, where the Br $K\alpha$ and P $K\alpha$ lines are dispersively detected, vs. the ratio of the actual number of bromine to phosphorus atoms in accurately prepared standards of simulated bromouracil replaced DNA is given in Fig. 15. A straight line approximates the data quite accurately. This indicates that the dispersive method described can be used to determine the ratio of the number of bromine to phosphorus atoms in 10 μg samples of DNA with an error which we estimate to be 5% for 10 min integration times with samples containing about 8% phosphorus and 0.8% bromine. Actual measures of the precision of the Br : P ratio have been given in Table IV. On occasion, we have found an inordinately large variance in the data of the bromine channel, when carrying out this type of precision analysis, which is decreased considerably by more careful mixing of the bromine

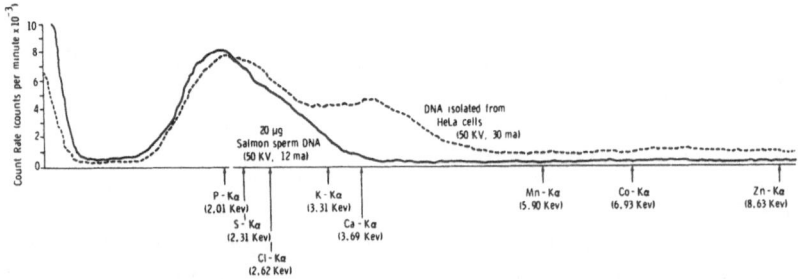

Fig. 14. X-ray nondispersive emission spectra of isolated and purified DNA from (i) salmon sperm and (ii) HeLa cells.

*This was kindly supplied by Dr. Jae Ho Kim.

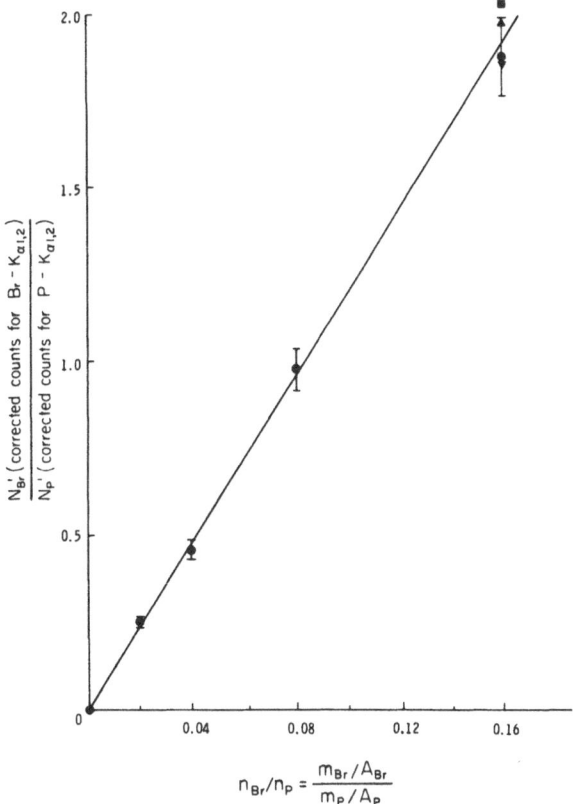

Fig. 15. Ratio of N'_{Br}/N'_P vs. the ratio of the number of atoms of bromine to atoms of phosphorus in accurately prepared, simulated standards of bromouracil replaced DNA. All samples are 10 µg.

containing DNA. The mixing must be extremely vigorous to obtain a homogeneous distribution of the added element in standards of DNA at a concentration of 1 mg/ml.

Three of the standards of Fig. 15 which contain 80 (Δ), 50 ($\square$), and 25% (∇) thymidine by weight simulate the case of determining the Br : P ratio in DNA samples with non-phosphorus containing impurities. The data for the simulated "impure DNA samples" indicate that this method allows for an accuracy of 5–6% in the determination of the Br : P ratio in samples with considerable nonphosphorus containing impurity where the bromine content is relatively high. The sample with 80% thymine has 2 µg of DNA with 3.3% bromine by weight and 8 µg of material simulating a nonphosphorus containing impurity.

With the resolution available in the vacuum spectrometer (see Fig. 11), it is possible to determine the degree of replacement of other analogs such as thioguanine, mercaptopurine, and chlorouracil by means of the detection of sulfur and chlorine without interferences from neighboring elements.

5. CONVENTIONAL EMISSION SPECTROSCOPY

The use of small sample techniques is limited to the analysis of elements where their concentrations are well above the parts per million level. Large sample analysis must be used to detect elements present at the level of a part per million or less. In our studies of zinc in human blood serum, following the approach of Gofman ([4]), we used large samples, 60 mg of briquetted, lyophilized serum. The dried powder discs, $\frac{3}{8}$-in. diameter, are produced by applying approximately 2000 lb/in.^{2} of pressure to the lyophilized serum powder placed in a Delrin lined stainless steel die with teflon discs above and below the dried serum. The die, Delrin liner, teflon discs, etc., as well as a completed disc are shown in Fig. 16. The discs are found to be very sturdy, surviving repeated exposure to X-radiation, and are stored in a microscope slide box in the manner shown in Fig. 16. Each disc is carefully weighed and all the data is corrected for variation in mass. The ratio of the mass of original serum to dried serum is also determined to allow for the conversion of the mass of zinc per unit mass of dry serum to the weight fraction in the liquid serum. The discs are placed on $\frac{1}{4}$-mil Mylar films used with the same configuration of specimen holder as shown in Fig. 1. The thicker material is easier to work with and there is no need

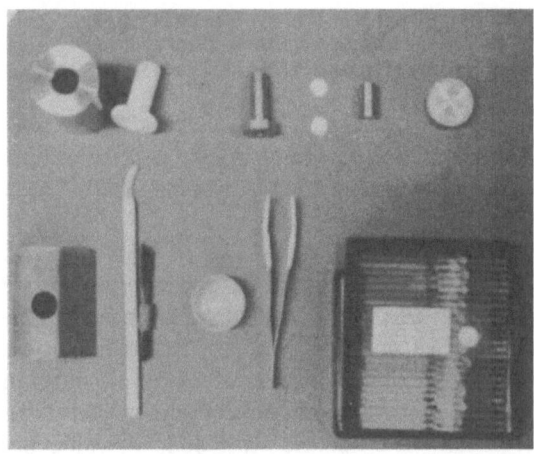

Fig. 16. Briquetting die, Delrin liner, Teflon discs, and accessories for producing dried powder discs. A briquetted human blood serum disc is shown, lower right, in well of plastic slide used as holder.

for the thinner nylon films since scattering from the film is now negligible compared to that from 60 mg of sample. The discs are carefully centered in the specimen holder with the aid of a positioning template. The beam of exciting X-rays is collimated so as to have a diameter at the plane of the sample which is slightly larger than that of the sample. Working curves for zinc in serum obtained by adding known amounts of zinc to the serum in the liquid state are found to be straight lines at least up to a concentration of added zinc which is 4 ppm wet weight fraction. The variance of the experimental points is found to be within the limits of confidence of counting statistics with this type of sample presentation.

It is simple to obtain relative values using this type of working curve, while much more difficult to make them absolute. Absolute determination requires either an independent measurement of one of the standards by another technique which is known to give accurate absolute concentrations or finding a material to determine the background level accurately. This material must have none of the element under study present, must scatter identically to the powdered blood serum ([4]) and must make a usable briquetted disc. We were unable to find such a material and based all our absolute determinations on a value of 1.00 ppm for the zinc level expected in the blood serum of normal humans ([4]). The values for zinc content used in the comparison to neutron activation analysis (Section 3.4.) were from working curves obtained in this manner.

The estimated limit of detectability in 60 mg dried blood serum is 0.4 ppm for bromine and 0.6 ppm for zinc (see Table III). The average concentration factor of blood serum, i.e., the ratio of the mass of the blood serum in the wet state to that in the dry, is found to be about 12. This means that the limit of detectability in parts per million of the original wet serum is about 0.03 ppm for bromine and 0.05 ppm for zinc.

6. COMPARISON OF SENSITIVITIES OF DISPERSIVE AND NONDISPERSIVE METHODS

A comparison of sensitivities of the dispersive and nondispersive methods for the detection of selected elements of interest was carried out in the following manner. Standards were chosen or prepared so as to contain a single heavier element, in relatively low concentration, in a light matrix such as sucrose. The signal intensity is measured at the position of the line of the element of interest in the sample with both dispersive and nondispersive spectrometers. The same is done for the background intensity measurement where the standard is replaced by a matrix sample of equal mass with none of the element of interest present. The line intensity is taken as the difference of the signal and background intensity.

Slit widths of the crystal spectrometers were chosen to give close to maximum sensitivity for all the lines studied. The proportional detectors in the non-dispersive instrument, either a proportional or scintillation counter, were used with filtration which we found to give optimum sensitivity (a maximum line squared over background, L^2/B) based on an empirical examination of various filter materials and thicknesses at different degrees of collimation. We found that a crystal spectrometer used with a proportional detector and pulse height selection showed the greatest sensitivity with no filtration. Therefore, the filtration used with the crystal spectrometer is only that which is an inherent part of the instrument, the beryllium window of the X-ray tube, beryllium or Mylar window of the specimen chamber and beryllium window of the detector. The pulse height analyzer (P.H.A.) was adjusted for maximum sensitivity for both the non-dispersive and dispersive instruments. The X-ray tube was maintained at 50 kV for the detection of all the elements. All the dispersive analyses were carried out at the maximum wattage at which our tubes are operated, i.e., 50 kV-30 mA for the Mo and W targets and 50 kV-20 mA for the Cr target tube.

There is a down-voltage line shift with increased integral count rate of about 1 % for every incremental 1000 integral counts/sec for the proportional counters* used in this study. In order to ensure an accurate analytical technique, the proportional counters are never used above 1000 integral counts per second when doing quantitative assays with the non-dispersive method. However, for this comparison study, we arbitrarily set 2500 counts/sec as the upper limit. In certain cases, the X-ray tube must be operated at less than the maximum permissible current value, with a tube voltage of 50 kV, in order to simultaneously satisfy the condition of filtration for optimum sensitivity and the imposed upper limit of integral count rate.

A scintillation counter gave greater sensitivity for the detection of Cadmium K (Cd K) radiation than our proportional counters. Therefore, it was used in this comparison study for the detection of Cd-K radiation even though its resolution is about two to three times inferior to a proportional counter. It is felt meaningful to use this detector since the object of this study is to compare sensitivities of the dispersive and nondispersive methods without regard for resolution. An integral count rate of 8500

---

*The proportional counter is an LND type 451B with a 0.002 in. diameter tungsten wire anode and 1 $\frac{15}{16}$-in. diameter aluminum cathode. The filling is 90% argon and 10% methane at 1 atm. The window is 0.001-in. beryllium. The pulses from the counter go to an Ortec 105XL preamplifier, followed by a Hamner N-340 linear amplifier and N-686 single-channel pulse height analyzer.

counts/sec is chosen as the upper level at which this detector and its associated electronics* should be used in an accurate analytical instrument.

The parameters used in carrying out this comparison are given in Table IX. It was found that in certain cases a different combination of detector collimation, tube current and filtration would give almost equal sensitivity to that obtained with the parameters shown in Table IX. For example, the sensitivity for the detection of bromine was the same as that shown with the following: (a) tube current, 30 mA; (b) detector collimation, 6.4×10^{-2} steradians; (c) X-ray tube filter, none; (d) detector filter, 0.0025 in. yttrium; and (e) integral count, 2400 counts/sec.

The results of the comparison are summarized in Table X. They indicate that for accurate quantitative analysis with the type of samples of interest in our work, even disregarding the inferior resolution of the non-dispersive method, there is no appreciable gain in sensitivity in the use of the nondispersive approach with a proportional counter in most of the energy region between 2–15 keV. There is a slight gain at the low energy end, 2.0 keV, with a proportional counter in a nondispersive instrument and an appreciable gain in sensitivity available above 15 keV with a proportional detector of high counting efficiency, such as, a scintillation counter or a lithium compensated silicon detector ([18]).

Any system of proportional detector plus its electronics, which allows the detector to be used at higher integral count rates than those specified above, will result in greater sensitivity for the nondispersive system. A lithium compensated silicon detector, if it can be operated at higher integral count rates without changes in line shape or line voltage, would result in considerably greater sensitivity for the nondispersive detection of lines whose energy is above 15 keV and in some improvement in the region from 2–15 keV. The comparative resolutions of the lithium drift detector and proportional or scintillation counter becomes more favorable to the solid state detector at higher energies. This means that, particularly at energies above 15 keV, the sensitivity of the solid state detector system can be improved in relation to the other detectors not only by the use of the detector with increased count rates, but also by the ability to increase the line to background ratio. The natural X-ray spectral linewidth is negligibly small compared to the resolution of the detectors under consideration. Therefore, the improved resolution allows for the detection of the line without loss of measured intensity with a smaller increment of spectrum detected on both sides of the line. If the background intensity is flat or approximately so in

---

*The scintillation counter is a Hamner SX10. The electronics beyond the preamplifier of the SX10 is identical to that given in footnote page 68. The SX10, with its 0.05-cm thick sodium iodide thallium activated phosphor, should have a photon counting efficiency of about 93% for Cd K X-ray photons ([18]).

TABLE IX

Proceedings of the 1967 Eastern Analytical Symposium,
L. Zeitz, X-ray Emission Analysis in Biological Specimens

	Phosphorus		Zinc		Bromine		Cadmium	
	N.D.	D.	N.D.	D.	N.D.	D.	N.D.	D.
X-ray[a] tubes — Target current, I_t	Cr 4 mA	Cr 20 mA	Mo 22.5 mA	Mo 30 mA	Mo 30 mA	Mo 30 mA	W 5 mA	W 30 mA
Filters { X-ray tube	none	none	0.0155" Al	none	none	none	0.00025"Ta	none
detector	none	none	none	none	none	none	0.0008" Ag	none
Crystal spectrometer — type		vacuum		air		air		air
crystal		PET[b]		LiF[b]		LiF		LiF
angular slit widths		$S_1 = 11'$ no S_2		$S_1 = S_2 = 13'$		$S_1 = S_2 = 13'$		$S_1 = S_2 = 16'$
%FWHM $(\Delta\lambda_{1/2}/\bar\lambda)$		0.30%		0.50%		0.90%		1.3%°
Detector — type	PC LND # 451B	PC LND (MAC)[d]	PC LND # 451B	SC SX10	PC LND # 451B ≃ 300	SC SX10	SC SX10	SC SX10
gas ampl. (A_G)	≃ 3000		≃ 300		≃ 300			
%FWHM $(\Delta E_{1/2}/E)$	34%	35% (est.)	16%	47% (est.)	12%	34% (est.)	25%	25%
collimation (steradians)	1.6×10^{-2}		1.6×10^{-2}		1.6×10^{-2}		6.4×10^{-2}	
P.H.A.—% $(\Delta V/\bar V)$	48%	71%	19%	71%	20%	71%	44%	71%
Integral counts/sec.	2,600	°	2500	°	2400	°	8500	°
% "down-voltage" line shift at above integ. c/sec.	2%		2%		2%		2%	

TABLE IX—Footnotes

[a] X-ray tube collimator is ¼ inch diameter titanium throughout. This results in a solid angle of 1·6 × 10⁻² steradians.

[b] PET—(002) planes of pentaerythritol; (200) planes of lithium fluoride.

[c] The resolution is actually inferior to that indicated by a 1·3% full width at half maximum. This results from the asymmetric line broadening, with tailing off of the line toward the high energy side, which occurs with higher energy photons due to Bragg reflections at greater depths in the crystal([18]).

[d] 0·001 inch Be window; 90% Xe, 10% CH_4; ¼ atmosphere.

[e] Integral count rate negligible compared to chosen maximum acceptable value.

TABLE X

Proceedings of the 1967 Eastern Analytical Symposium, L. Zeitz, X-ray Emission Analysis in Biological Specimens

Sample mass αg	Phosphorus		Zinc		Bromine		Cadmium	
	N.D.	D.	N.D.	D.	N.D.	D.	N.D.	D.
Line	P–K 2.0 keV	P–Kα$_{1,2}$ 6.155 Å	Zn–K 8.6 keV	Zn–Kα$_{1,2}$ 1.437 Å	Br–K 11.9 keV	Br–Kα$_{1,2}$ 4.041 Å	Cd–K 23.2 keV	Cd–Kα$_{1,2}$ 0.536 Å
N_L[a]　512	d	d	8.8×10^4	4.7×10^3		1.3×10^4	1.9×10^5	2.3×10^3
10	7.5×10^4	3.2×10^2	[e]3.4×10^5	4.7×10^3	1.8×10^5	1.3×10^4	[e]9.4×10^5	2.3×10^3
N_B[b]　512			1.3×10^4	81		3.1×10^2	2.3×10^5	9.9×10^2
10	2.4×10^4	3	[e]1.8×10^4	6	5.5×10^3	23	[f]1.5×10^5	1.4×10^2
L.D.[c]　512	3×10^{-9}		2×10^{-9}	2×10^{-9}		2×10^{-9}	3×10^{-9}	2×10^{-9}
10	7×10^{-9}		5×10^{-10}	7×10^{-10}	5×10^{-10}	5×10^{-10}	5×10^{-10}	7×10^{-9}

[a] N_L—counts/minute/μg element.

[b] N_B—counts/minute for matrix plus film with none of the element of interest present.

[c] L.D.—Limit of detectability for 10 minute integration. See Section 3.4 for definition.

[d] There is appreciable self-absorption of P–K lines in 512 μg samples.

[e] The maximum acceptable integral count rate is obtained for the 10 μg samples with no filtration on the X-ray tube or detector and I_t of 30 mA.

[f] The maximum acceptable count rate is obtained for the 10 μg sample with the same filtration as the 512 μg sample and I_t of 30 mA.

the vicinity of the line, then a decrease in the spectral increment detected will result in a decreased measured background. If the predominant background in the measured signal has the same spectral linewidth as the line of interest and is at the position of the line of interest, or very close to it in comparison to the resolution of the detector; then there is little to be gained in improvement of the line to background ratio by the use of a detector with superior resolution.

ACKNOWLEDGMENTS

This work was supported in part by Public Health Research Grant HD01410 and NCI grant CA08748 from the National Institutes of Health.

The author wishes to thank Mr. Richard Lee for his valuable technical assistance.

The author also thanks the editors of *Analytical Biochemistry* and the *American Journal of Physiology*, who kindly allowed the author to draw upon material previously published by him in these journals.

REFERENCES

1. L. Zeitz and R. Lee, Zinc Analysis in Biological Specimens by X-ray Fluorescence, *Anal. Bioch.* **14**, 191–204 (1966).
2. L. Zeitz and R. Lee, Element Analysis in Labeled DNA by X-ray Fluorescence, *Anal. Bioch.* **23**, 442-458 (1968).
3. H. A. Liebhafsky, H. G. Pfeiffer, E. H. Winslow, and P. D. Zernany, *X-ray Absorption and Emission in Analytical Chemistry*, John Wiley and Sons, New York (1960).
4. J. W. Gofman, in *Advances in Biology and Medical Physics* (C. A. Tobias and J. H. Lawrence, eds.) Vol. 8, Academic Press (1962) pp. 1–39.
5. S. Natelson, M. R. Richelson, B. Shied, and S. L. Bender, X-ray Spectroscopy in the Clinical Laboratory, *Clin. Chem.* **5**, 519–531 (1959).
6. S. Natelson, X-ray Spectrometric Determination of Strontium in Human Serum and Bone, *Anal. Chem.*, **33**, 396–401 (1961).
7. P. K. Lund, D. A. Morningstar, and J. C. Mathies, in *X-ray Optics and X-ray Microanalysis*, (H. H. Pattee, V. E. Cosslett, and A. Engström, eds.), Academic Press (1963).
8. T. Hall, X-ray Fluorescence Analysis in Biology, *Science*, **134**, 449–455 (1961).
9. L. Zeitz and R. Lee, Bromine Analysis in 5-bromouracil Labeled DNA by X-ray Fluorescence, *Science* **142**, 1670–1673 (1963).
10. T. Johansson, A New Exactly Focusing Röntgen Spectrometer, *Z. Physik* **82**, 507–528 (1933).
11. L. Zeitz, The Design of an X-ray Microprobe Analyzer for Biological Specimens, Biophysics Laboratory Report No. 67 (1962).
12. A. H. Compton and S. K. Allison, "X-rays in Theory and Experiment," D. Van Nostrand Co., New York, Second edition (1954) p. 90.
13. A. Engström, Note on the Cytochemical Analysis of Elements by Roentgen Rays, *Acta Radiol.* **36**, 393–396 (1951).

14. L. Zeitz, X-ray Fluorescence Analysis in Biological Specimens, Microfilm file #AED-Conf 1966/353/3, AEC Depository Library at the Gmelin Institute, Frankfurt, Germany.
15. S. Saito, L. Zeitz, I. M. Bush, R. Lee, and W. F. Whitmore, Jr., Zinc Uptake and EDTA—Deprivation in Canine or Rat Spermatozoa, *Amer. J. Physiol.* 217 (1969).
16. S. Saito, L. Zeitz, I. M. Bush, R. Lee, and W. F. Whitmore, Jr., Zinc Content of Spermatozoa From Various Levels of the Canine and Rat reproductive Tract, *Amer. J. Physiol.* 213, 749–752 (1967).
17. J. I. Trombka, I. Adler, R. Schadebeck, and R. Lamother, Nondispersive X-ray Emission Analysis for Lunar Surface Geochemical Exploration, Goddard Space Flight Center Report #X-641-66-344 (Aug. 1966).
18. J. Taylor and W. Parish, Absorption and Counting-Efficiency Data for X-ray Detectors, *Rev. of Sci. Instr.* 26, 367–373 (1955).
19. L. Alexander, A Synthesis of X-ray Spectrometer Line Profiles with Application to Crystallite Size Measurements, *J. Appl. Phys.* 25, 155–161 (1954).
20. H. R. Bowman, E. K. Hyde, S. G. Thompson, and R. C. Jared, Application of High Resolution Semiconductor Detectors in X-ray Emission Spectrography, *Science* 151, 562–568 (1966).

V. THE APPLICATION OF ELECTRON PROBE MICROANALYSIS TO THE STUDY OF AMINO ACID TRANSPORT IN THE SMALL INTESTINE*

A. J. Tousimis, Jon C. Hagerty, and
Thomas R. Padden

*Biodynamics Research Corporation,
Rockville, Maryland*

and

Leonard Laster

*Section on Gastroenterology, Metabolic Diseases Branch,
National Institute of Arthritis and Metabolic Diseases,
Bethesda, Maryland*

The purpose of this investigation was to explore the utilization of electron probe microanalysis for the study of solute transport by epithelial cells. Segments of small intestine from golden hamster or *Amphiuma means* were incubated in buffer containing 40 mM DL-selenomethionine, removed at various times between 0 and 16 min, and processed for electron probe analysis. The accumulation of the amino acid in the epithelial cells lining the intestinal specimens was assessed by measurement of X-ray intensity of Se $L\alpha$. The results suggest that selenomethionine content increased with time and reached a maximum value at approximately 4–8 min. When the experiment was repeated with 0.2 mM 2,4-dinitrophenol in the incubation medium, accumulation of selenomethionine appeared to be partially reduced. The results suggest that the probe analyses reflected an uptake process and the inhibition by 2,4-dinitrophenol suggests that at least part of this process was active transport. These preliminary observations provide evidence that electron probe microanalysis may eventually prove to be a useful tool in studying biological phenomena such as membrane transport. Difficulties encountered in tissue preparation and in attempts to repeat some of the observations, all emphasize the need for additional technical development.

*This research was supported by the National Institute of Arthritis and Metabolic Diseases, National Institutes of Health under Contract No. PH-43-66-10.

1. INTRODUCTION

The general application of electron probe microanalysis to the study of biological tissues will be an important technological advance that will enable investigators to localize and quantitate elements at the subcellular level. At present, the analytical procedure, and methods for preparing tissues for the analysis, require extensive further development. In addition to determining the elemental content of various tissues, it would be interesting to examine dynamic events such as changes in elemental composition of a tissue during *in vitro* experiments. As an approach to this problem, we explored the feasibility of applying electron probe microanalysis to the study of the transport of solutes across epithelial surfaces. In this paper, we report preliminary results of an investigation of the *in vitro* movement of the amino acid, DL-selenomethionine, across the epithelium of the small intestine of the golden hamster and of *Amphiuma means* (Congo eel). The studies were directed toward delineating and defining the problems to be encountered.

2. MATERIALS AND METHODS

2.1. Experimental Animals

The golden hamster was selected because the size of the epithelial cells of its small intestine is similar to that of the human, approximately $5 \times 15 \mu$. *A. means* was selected because its intestinal epithelial cells are approximately $9 \times 60 \mu$, and this large cell size was helpful in the development of our experimental procedures. The hamsters were obtained from an animal colony at the National Institutes of Health, the specimens of *A. means* were caught in Lousiana.*

The hamster small intestine is approximately 30 cm long and 0.5 cm in diameter. Finger-like villi project into its lumen from the gut wall, Fig. 1. The villi vary somewhat in length but on the average they are approximately 500μ long and $125 - 175 \mu$ wide. The epithelium covering the villi comprises absorptive columnar cells and secretory goblet cells. In the region of the hamster small intestine that we studied, the ratio of columnar to goblet cells on the villus is approximately nine to one.

The *A. means* small intestine does not have true villi. The luminal surface is thrown into deep folds, or rugae, Fig. 2, which are analogous to villi in that they present a large epithelial surface area to the luminal contents. The size of the rugae varies along the length of the intestine. The ratio

*Purchased from the Carolina Biological Supply Company, Burlington, North Carolina.

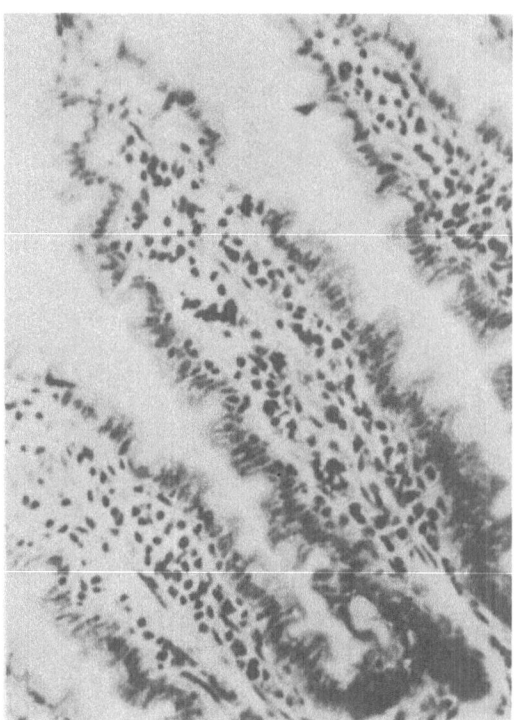

Fig. 1. Light micrograph of a longitudinal section of villi from hamster small intestine; 5 μ thick, paraffin-embedded section, PAS stain. 360X.

of columnar to goblet cells is not constant, even within one region of the intestine, and appears to vary with the animal's size, activity, or both. Rugae projecting almost to the center of the lumen are frequent and may have dimensions in cross section, as large as 2000 by 250 μ.

2.2. Experimental Procedure

The plan of our experiments was to incubate specimens of small intestine in the presence of an amino acid, remove the tissues at various time intervals, and determine the amount of one of the elements of the amino acid in the intestinal epithelial cells. The elements in the conventional amino acids, carbon, nitrogen, oxygen, hydrogen, and sulfur, are present in tissues in such abundance that one cannot expect to detect changes in their amounts secondary to the accumulation of an amino acid. Therefore, we elected to work with selenomethionine, an analogue of methionine with selenium in place of sulfur, Fig. 3. Preliminary electron probe analyses failed to reveal any selenium in the normal hamster small intestine. Seleno-methionine has been shown to be transported against a concentration gradient by hamster small intestine and it appears to be handled identically

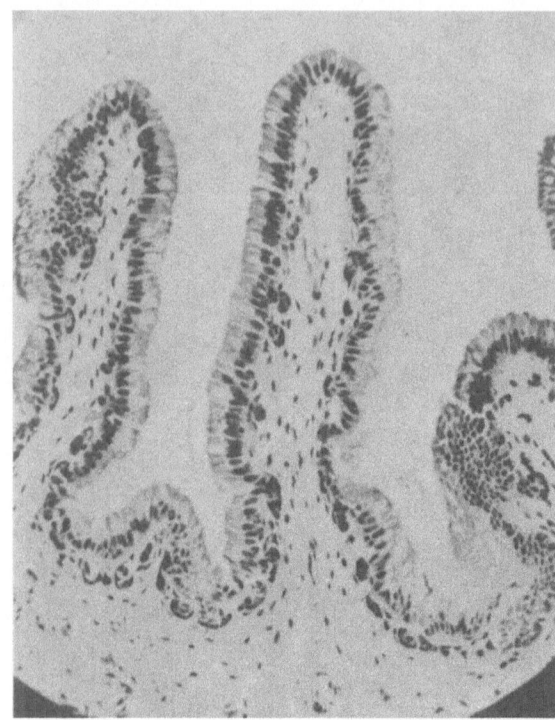

Fig. 2. Light micrograph of a longitudinal section of *A. means* small intestine; 5 μ thick, paraffin-embedded section, H & E. stain. 96X.

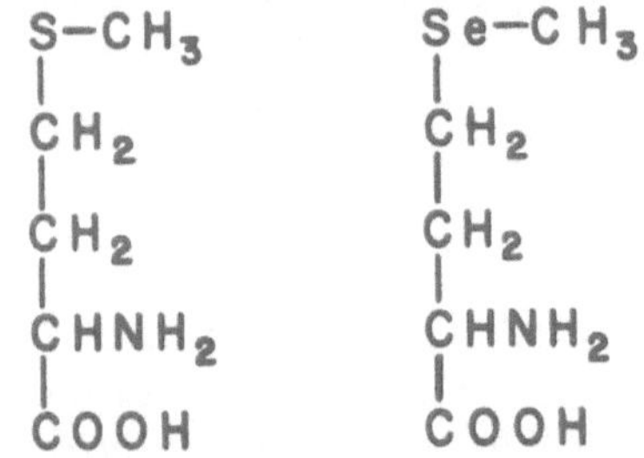

$$S-CH_3 \qquad Se-CH_3$$
$$CH_2 \qquad\quad CH_2$$
$$CH_2 \qquad\quad CH_2$$
$$CHNH_2 \qquad CHNH_2$$
$$COOH \qquad COOH$$

Methionine   Selenomethionine

Fig. 3. Structural formulas for methionine and selenomethionine.

to methionine ([1]). Because intestinal transport systems exhibit a marked specificity toward the L-form of amino acids rather than the D-, ideally we would have preferred to use L-selenomethionine for our studies; however, only the DL-form was available to us. It was a commercial product.*

*Purchased from Cyclo Chemical Corporation, Los Angeles, California.

Intestinal segments were removed from the distal third of the hamster small intestine, the region in which amino acid transport appears to be most active ([2]), opened longitudinally, and cut into rectangles approximately 0.5 by 1.0 cm. The specimens were washed in, and pre-incubated for 2 min at 37°C in Krebs–Ringer phosphate buffer containing 200 mg/ 100 ml glucose. Then they were transferred to the same medium containing in addition 40 mM DL-selenomethionine, and incubated at 37°C. Solutions were oxygenated for 5 min before and throughout the experiment. Specimens were removed for analysis at the indicated times and processed as described in Section 2.3. Lacking information about transport in *A. means* small intestine, we selected no particular region of the bowel. Segments approximately 1.0–1.5 cm square were taken from the intestinal wall and preincubated and incubated as described for hamster gut except that the temperature was 22°C and at the conclusion of the incubation period tissue specimens were dipped in buffer containing 3% gelatin and 40 mM DL-selenomethionine just before freezing. The specimen for zero time was dipped in buffer containing only 3% gelatin.

The effect of 2,4-dinitrophenol was studied with *A. means* small intestine. It was present in the medium during both the pre-incubation and the incubation periods at a concentration of 0.2 mmole/liter.

2.3. Electron Probe Microanalysis

Electron probe analysis requires a specimen with electrical conductance and with sufficient thermal stability to withstand irradiation by an electron beam in a vacuum without loss of the elements under analysis. Specimens meeting these requirements have been prepared from various biological tissues including intestine ([3]). Many of the earlier methods were used in the present study and additional procedures were also developed.

Electron probe analysis of molecules and ions, presumed to be freely diffusible within a tissue, necessitates some restrictions on sample preparation not encountered when investigating bound molecules or ions. If a tissue section is dried in air, unbound molecules or ions may diffuse within the tissue during evaporation of water, and as the water–air interface passes through the tissue structure there may be a sweeping action at the interface with a resultant redistribution of the unbound molecules and ions. To avoid this potential difficulty, the specimens used in the present study were prepared by freeze-drying as follows.

After incubation, specimens were frozen rapidly at −150 to −160°C in isopentane cooled in liquid nitrogen. Sections 6–10 μ thick were cut from the frozen tissue in a cryostat microtome, transferred to a block cooled in liquid nitrogen, and vacuum-desiccated to sublime the ice without permitting it to pass through a liquid phase. The desiccated sections were mounted

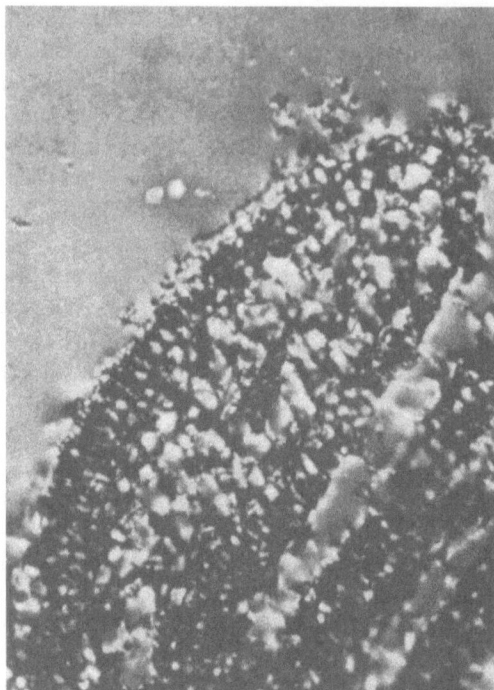

Fig. 4. Light micrograph of a frozen-dried longitudinal section of a villus from hamster small intestine mounted with epoxy on an epon disc and coated with aluminum for electron probe analysis. 360X.

on epoxy resin discs by use of a very thin layer of partially polymerized viscous epoxy resin as a glue, care being exercised not to impregnate the tissue during mounting. After final curing of the epoxy resin, the specimens were coated with either carbon or aluminum to achieve the electrical conductivity needed for the electron probe analysis.

Figure 4 is a micrograph of hamster intestine prepared in this way. Unfortunately, subcellular structure is not readily discernible. Although the specimen preparation procedure was not completely satisfactory for microscopy, it was adequate, when appropriate procedures were used to measure X-ray intensity, to determine whether the uptake of selenomethionine by intestinal columnar cells can be detected by the electron probe. The specimens were analyzed with an electron probe* operated at 15 keV and 0.015-μA specimen current. Under these conditions 2370 counts/sec K $K\alpha$ intensity was measured from a KCl standard (LiF dispersive crystal in the vacuum spectrometer) with a 1.2 counts/sec background measurement on either side of the potassium line; 700 counts/sec Se $L\alpha$ intensity

*Materials Analysis Company, Palo Alto, California, electron microprobe No. 400.

was recorded from the selenium standard (pentaerythritol crystal) with a background measurement of less than 1 count/sec.

We attempted to select villi for analysis, by use of preliminary procedures to be described in another communication, so that the villi were of approximately uniform and comparable thickness. The probe beam was located in the region of the epithelium midway between the nucleus and the brush border of the cells. The entire epithelial layer of a villus was analyzed, point by point, by moving the specimen so as to advance the beam of the probe 3 μ at a time for hamster intestine and 5 μ at a time for *A. means* intestine. A period of 30 sec was used at each point to measure the intensity of Se $L\alpha$ or, in some instances, K $K\alpha$. Each epithelial cell was analyzed at least once. The values obtained for all the sites on a villus were averaged and in the case of selenium, the value so obtained was regarded as an index of uptake of selenomethionine. We recognize that goblet cells, which presumably do not take part in the transport process, were included in the measurements. The track left by the electron beam during analysis of a hamster villus is depicted in Fig. 5. In some specimens there were obvious cracks, perhaps the result of our tissue preparation procedure,

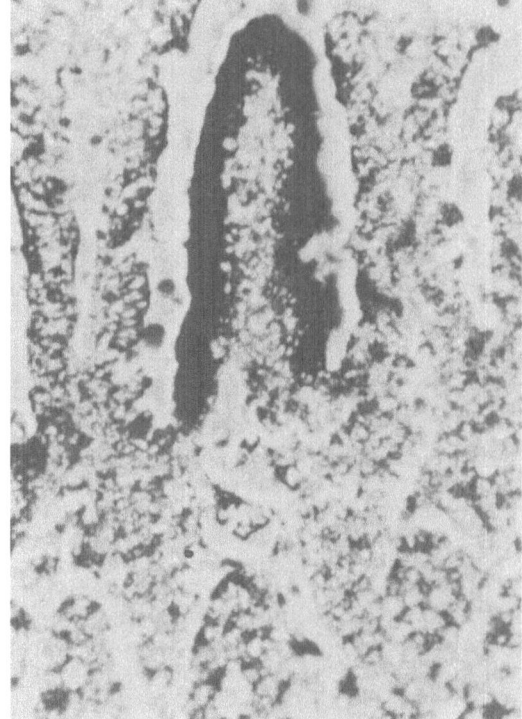

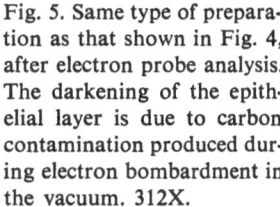

Fig. 5. Same type of preparation as that shown in Fig. 4, after electron probe analysis. The darkening of the epithelial layer is due to carbon contamination produced during electron bombardment in the vacuum. 312X.

and when these passed through the epithelium of the villus under analysis, the area in which they occurred was skipped.

3. RESULTS

3.1. Evidence for Uptake of Selenomethionine

Incubation of hamster small intestine in medium containing DL-selenomethionine resulted in an apparent increase in the selenium content of the epithelium between 1 and 4 min, without further change between 4 and 16 min, Fig. 6. The variability encountered in analyzing the specimens, especially those incubated for 4 min, may be due, in part, to failure to select villi of comparable and uniform thickness. This problem is presently under study.

Data for determinations of Se $L\alpha$ and K $K\alpha$ at each point examined on an individual villus are shown in the graph in Fig. 7. The variations in

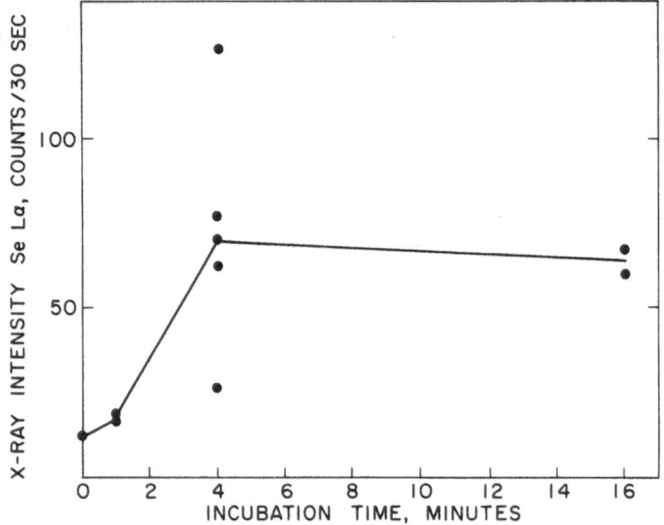

Fig. 6. Results of electron probe microanalysis of selenium in epithelial cells of hamster small intestine plotted as a function of duration of exposure of tissue to DL-selenomethionine. The intestinal segments were incubated under conditions described in the text. Two villi were analyzed for the 1-min specimen, five for the 4-min specimen, and two for the 16-min specimen. The number of sites analyzed ranged from 80 to 230 per villus. Selenium content is expressed as X-ray intensity of Se $L\alpha$ in counts per 30 sec. Circles represent mean values for individual villi. The solid line connects the average of the means for the individual villi.

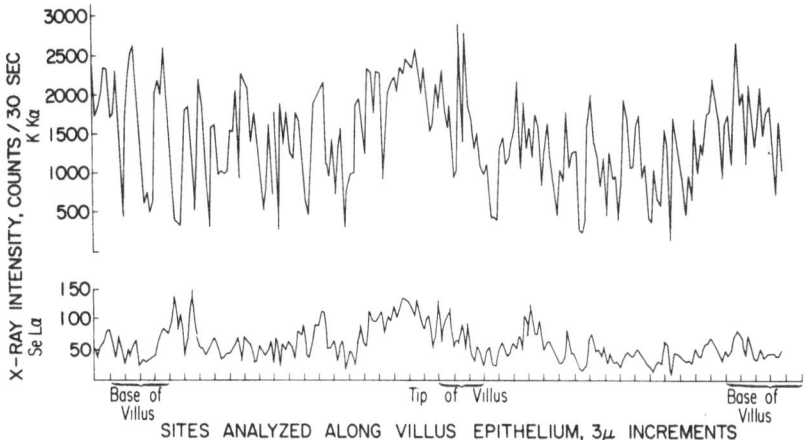

Fig. 7. X-ray intensities of selenium (Se $L\alpha$) and potassium (K $K\alpha$), measured simultaneously from the apical region of epithelial cells of a single villus from hamster small intestine and plotted as a function of site on villus. Analyses were begun at the base of one side of the villus and the probe was advanced at 3 μ increments around the periphery of the entire villus. The specimen was exposed to DL-selenomethionine for 16 min. Additional details in text and in legend to Fig. 6.

intensity are obvious. We have not answered such questions as whether these variations reflect differences in transport activity from cell to cell around the villus or whether they are technical in origin. Another question to be studied is whether there is a negative or positive correlation between the variations in selenium and potassium.

When intestinal specimens from *A. means* were incubated in medium containing DL-selenomethionine, the selenium content of the epithelium increased in a pattern comparable to that observed for the hamster (Fig. 8, circles and solid line), except that the marked increase in selenium appeared to occur between 2 and 8 min.

3.2. Effect of 2,4-Dinitrophenol

Based on what is known about the structural requirements for active transport by the small intestine, one would expect that only the L- form of DL-selenomethionine would undergo active transport. The D-form would be expected to diffuse across the gut epithelium. If any of the apparent accumulation of selenium suggested by the data in Fig. 8 were attributable to active transport of selenomethionine, one would expect, too, that inhibitors of active transport would partially reduce the accumulation. To explore whether the electron probe analysis was indeed providing information related to active transport, we repeated the study of *A. means*

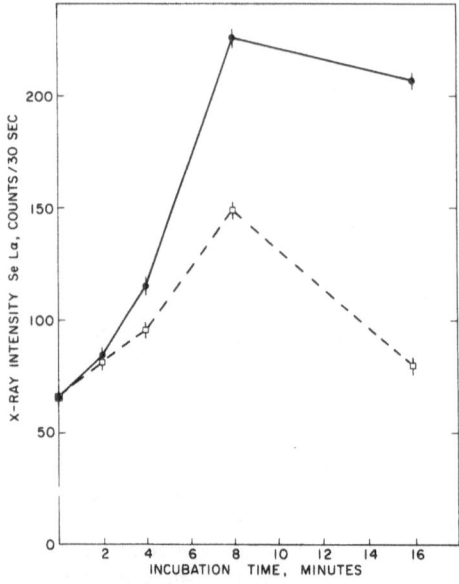

Fig. 8. Results of electron probe micro-analysis of selenium in epithelial cells of *A. means* small intestine plotted as a function of duration of exposure of tissue to DL-selenomethionine. For additional details see Fig. 6 and text. ●, 40 mM DL-selenomethionine in incubation medium; □, same plus 0.2 mM 2,4-dinitrophenol. Vertical lines from each symbol represent the standard error of the mean. One villus each was analyzed for the 0-, 1-, and 16-min specimens, two villi each for the 4- and 8-min specimens.

intestine with 2,4-dinitrophenol in the incubation medium at a concentration known to inhibit active transport of amino acids by small intestine from other animals. The results (Fig. 8, boxes and broken line) strongly suggest that uptake of selenomethionine, as assessed by electron probe microanalysis, was partially inhibited.

4. DISCUSSION AND CONCLUSIONS

Although, at present, we feel that the data presented in Figs. 6–8 are not sufficiently reliable to permit definite conclusions, they do suggest that the electron probe determinations of Se $L\alpha$ intensities in the intestinal epithelial cells after exposure of the tissues to DL-selenomethionine reflect uptake of the amino acid by the epithelial cells from the incubation medium. The apparent inhibition by 0.2 mM 2,4-dinitrophenol is consistent with the interpretation that some of the uptake was mediated by an active transport system, since 2,4-dinitrophenol is well known to inhibit active transport of amino acids by the intestinal mucosa. These preliminary observations indicate the need for continued exploration of this subject and for improvement in the experimental procedures.

A major requirement is an improved method for specimen preparation, one that will retain the advantages implicit in freeze-drying but will also preserve morphological boundaries and thereby enable the investigator to localize elements to specific locations and organelles within the epithelial cells. In addition, we need a method for assessing specimen thickness at the site of determination of X-ray intensities and at the time that these intensities are measured. Such improvements would be expected to reduce the variation in results attributable to technical factors and this would permit an evaluation of variations due to biological phenomena.

The intensities of Se $L\alpha$ that were encountered in the present experiments were very low, and indeed when we performed additional experiments, comparable to the ones reported here, the observed intensities over the tissues were too low to differentiate from background values. This problem might have been due to biological factors. Thus, the hamsters that we used for the later experiments were found to have been infected with *Giardia lamblia* which might have interfered with selenomethionine uptake. When the repeat experiments with *A. means* failed, we realized then that the animals were much more sluggish than they had been previously. We learned that this animal goes into an inactive or "dormant" state during certain seasons of the year, even when maintained at laboratory temperatures, and it is conceivable that this dormancy in some way was responsible for the failure to detect uptake of selenomethionine by the intestinal segments in the repeat experiments. However, because of uncertainties about the technical aspects of the study, we cannot exclude the possibility that for reasons as yet not clear, the analytical procedure was responsible for the difficulties.

We conclude that our preliminary findings are encouraging, and additional studies are now in progress.

REFERENCES

1. R. P. Spencer, and M. Blau, *Science* **136**, 155 (1962).
2. D. M. Matthews, and L. Laster, *Am. J. Physiol.* **208**, 593 (1965).
3. A. J. Tousimis, *Biomed. Sci. Instr.* **1**, 249 (1963).

VI. ELECTRON PROBE MICROANALYSIS OF BIOLOGICAL STRUCTURES*

A. J. Tousimis

Biodynamics Research Corporation
Rockville, Maryland

The historical background and principles of electron probe microanalysis are presented. The effects of electron bombardment specimens and general procedures for specimen preparation are reviewed. The biological applications of electron probe analysis are briefly discussed.

1. INTRODUCTION

1.1. Principle of the Instrument

A high-efficiency electron gun (Fig. 1) is used to produce electron beams with energies between 1 and 30 KeV. By the double condenser lens together with an objective lens located above a horizontally or obliquely placed specimen, the electron beam is reduced to $0.1\,\mu$ or up to $1\,\mu$ in diameter with electron fluxes from $0.5\,\mu A$ to a few nanoamperes. The electron beam current falling onto the specimen is controlled by varying the focal length of the condenser lens over an aperture located just below it. The beam can be positioned or scanned over the specimen via electromagnetic scanning coils located just above the objective lens. The primary or secondary electrons (usually less than 50 eV) backscattered from the specimen as the electron beam (probe) interacts with it are measured with electron detectors located above the specimen or in the case of ultrathin sections below it.

Secondary electrons are collected and accelerated by a biased gold screen kept at 300 V positive in front of a hemispherical aluminum coated scintillator kept at 10 KeV. The light generated at the scintillator after the post-accelerated secondary electrons strike it, is relayed through a

*This research was supported in part by the National Institutes of Health under grant AM-06350 and contracts Nos. PH-43-66-10 and PH-44-67-104.

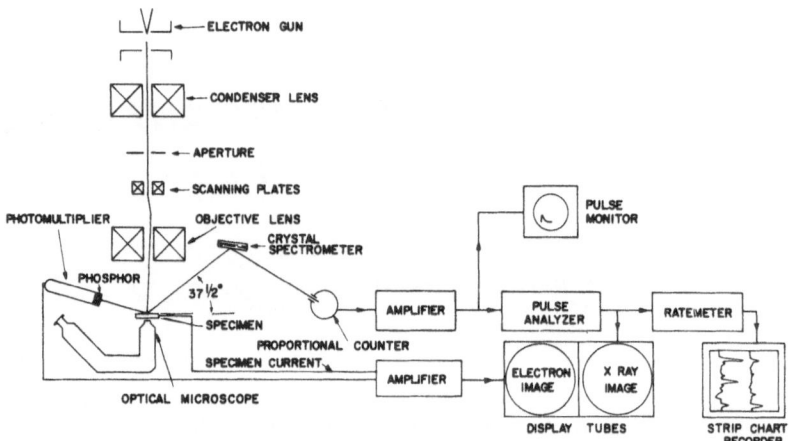

Fig. 1. Schematic diagram of the Castaing-type configurations showing apertures of lenses used in the electron probe. Electrons originating from the hot tungsten filament of an RCA electron gun are accelerated by potentials varying from 1–30 KeV. The lenses are electromagnetic. The molybdenum aperture is placed between the double condenser and objective lenses. Opaque specimens are observed with a reflective glass lens (not shown) located inside the specimen chamber. Thin sections of biological specimens are observed from below. In the arrangement depicted schematically in this figure, X-rays emitted by the electron bombarded specimen are observed by a crystal spectrometer. The angle between the surface of the specimen and the line of sight of the diffraction crystal or detector (in the nondispersive cases) is more than 30°. Backscattered electrons, transmitted electrons, and specimen currents as well as the proportional counter outputs can be recorded on display tubes (oscilloscope) or on a strip-chart recorder and scalers (not shown). For additional details see text.

light pipe to a photomultiplier; this signal is amplified and can be made to modulate a synchronously scanned oscilloscope electron beam (display tube) producing a scanning electron micrograph that is characteristic of the surface morphology of the specimen. Even though the same type of electron detector can be used to detect higher-energy backscattered electrons, a solid-state type such as silicon diodes are preferred. The output of this detector also modulates the intensity of the display tube resulting in a scanning electron micrograph containing more information—in depth of the specimen—than that obtained using secondary electrons.

The X-rays which are emitted from the electron bombarded specimen contain chemical information in terms of characteristic lines of the elements. The intensities of the characteristic lines are measured with suitable X-ray detectors before or after they are dispersed by a diffraction crystal. X-ray lines with wavelengths from less than 1 Å to more than 100 Å can be measured with sufficient signal-to-noise ratio for analysis of amounts lower than 0.1% in a cubic micron of specimen, or 10^{-15}–10^{-16} g absolute amounts. Similar to the electron scanning micrographs discussed above, micrographs of the specimen are obtained in terms of an element, showing the improved spatial distribution and intensity concentration variations from this method. An instrument of this construction can be used to analyze biological specimens.

1.2. History and Background of the Method

At this point it might be appropriate to say something about the early beginnings of this methodology. In 1914, Moseley in his classical experiments ([1]) used a trolley carrying a number of different samples which were exposed to an electron beam. Application of the principle of analysis by direct electron excitation of specimen has since then been slow in its exploitation, in contrast to the X-ray excitation procedures which have become very highly refined and are in common use.

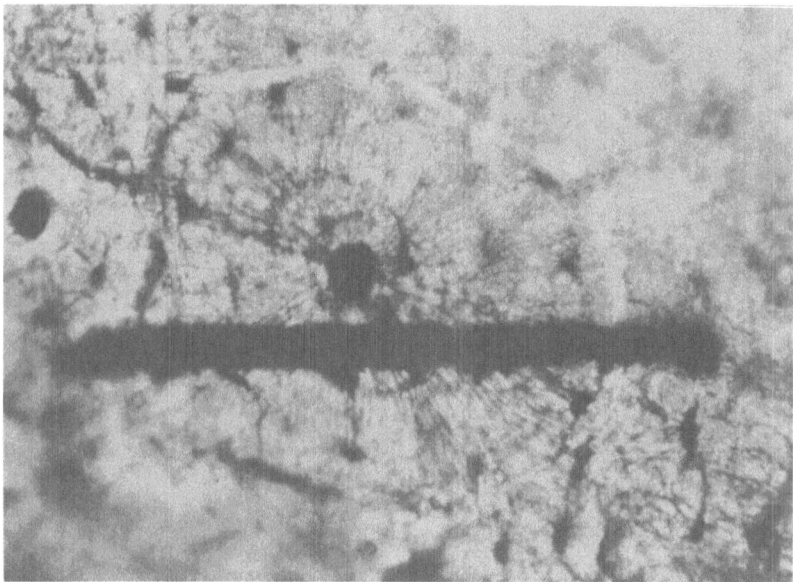

Fig. 2. One of the first biological specimens to be examined with a stationary spot electron probe. The specimen is a 40 μ thick section of human long bone. The dark line is the contamination left on the surface of the specimen just below a Haversian canal as the specimen was mechanically moved under the stationary electron beam and the calcium characteristic X-ray line intensity was recorded on a strip-chart recorder.

A major impediment in developing electron probes has been the difficulty in controlling the energy and intensity of electron beams with the necessary precision for analysis. In addition, most early attempts with relatively large electron beams in elemental analysis were not successful because of severe damage to the specimen before meaningful analysis could be made. Otherwise, this would have been the choice of elemental analysis, since the excitation by electrons is more efficient than by X-rays. In 1943, Hillier [2] proposed using the electron probe to carry out chemical analysis on opaque specimens. Subsequently, Castaing [3] in France, Borovskii [4] in Russia, Birks and Brooks [5] of the Naval Research Laboratories, and others constructed electron probes and successfully applied them to some metallurgical and mineralogical problems. In 1957 Tousimis, Birks, and Brooks [6] applied the electron probe to biological specimens. In these studies the calcium distribution in sections of mature bone was determined. A light micrograph of the first bone specimen to be analyzed is shown in Fig. 2. At that time it was shown that the variation of an element in a

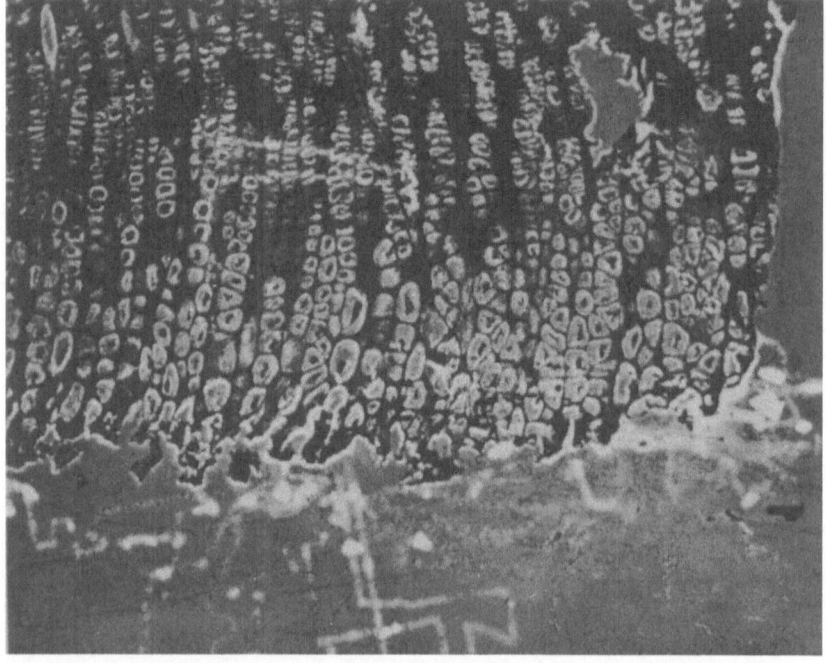

Fig. 3. Light micrograph of 2 μ thick epiphyseal cartilage section. After the section was placed on the quartz surface, the methacrylate embedding medium was dissolved with xylene. Light line tracks across the section and on the quartz surface are portions of specimen exposed to the electron beam. The specimen is coated with aluminum.

Fig. 4. Objective lens of the electron optical column with the specimen stage and light microscope attachment of a laboratory-built electron probe system.

biological specimen could be traced and that the specimen portion an-alyzed was not damaged by the electron beam during analysis and could be retained for subsequent study. Thin sections of cartilage (Fig. 3) were also analyzed for calcium ([7]). Progress with thin sections of specimens sug-gested the possible improvements in spatial resolution in electron probe analysis. During the last ten years many improvements in both specimen preparation procedures and probe instrumentation as well as X-ray intensity data processing in the study of biological tissues have taken place. Selected events are presented in this symposium commemorating the tenth anniversary since the first application of the electron probe to biology.

Fig. 5. The electron optical system with the electron gun on the top, centerable aperture and final lens. The operator is centering the specimen with the light microscope attachment on the specimen stage.

2. RECENT INSTRUMENTATION

2.1. Resolution

The resolving power of the electron probe is limited by the diameter of the electron beam impinging on the specimen, the chemical composition and density of the specimen, the thickness of the specimen, and the sensitivity of the detector system with its associated electronic components. While the primary function of the standard electron optical column is to produce an

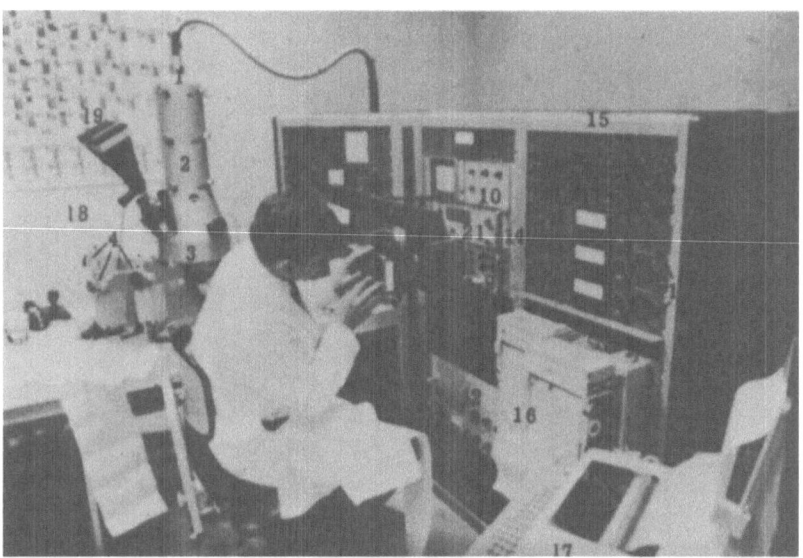

Fig. 6. The combined scanning electron microscope-electron probe used for micro-analysis of biological tissues. (1) Electron Gun Assembly; (2) Double Condenser Lens; (3) Objective Lens; (4) Spectrometers—housing analyzing crystals, detectors, and pre-amplifiers; (5) High Voltage Cage; (6) Lens Power Supply is below High Voltage Cage and is obscured by operator; (7) Step Scanning System located below lens power supply is also obscured by operator; (8) Pico-ammeter—Specimen Current Readout; (9) Electron Beam Scanning System—Controls and Oscilloscope Readout, Oscilloscope obscured by oscilloscope camera; (10) Electron Beam Scanning System Auxiliary Scope; (11) X-ray Channel Selection Panel; (12) High Voltage Power Supply for Detectors; (13) Pulse Height Discriminators and Linear Amplifiers for each detector obscured by strip chart paper; (14) Ratemeters for each counting system; (14a) Rate-meter Audio Output; (15) Scalers for each counting system; (16) Dual Channel Recorders record each ratemeter output and specimen current simultaneously; (17) Teletype Printout, six channel capacity; (18) Light Optics Microscope; (19) Micro-scope Camera; (20) High Vacuum System Controls; (21) Digital Voltmeter for Specimen Current Readout; (22) Oscilloscope Camera.

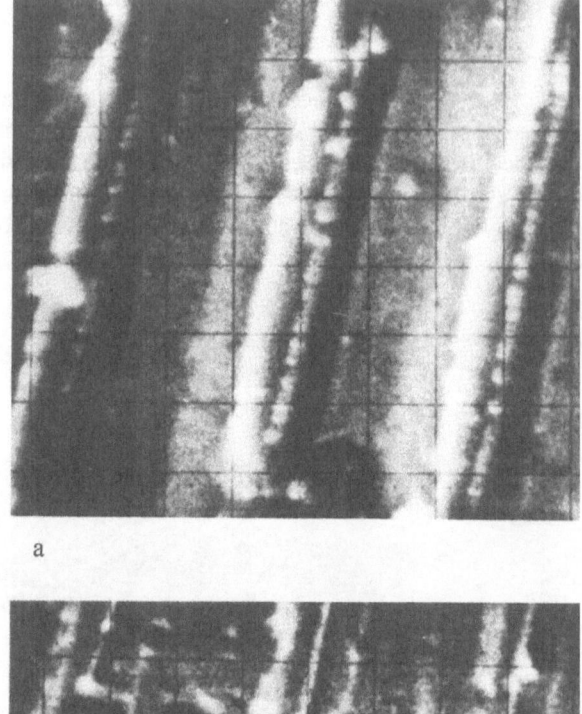

a

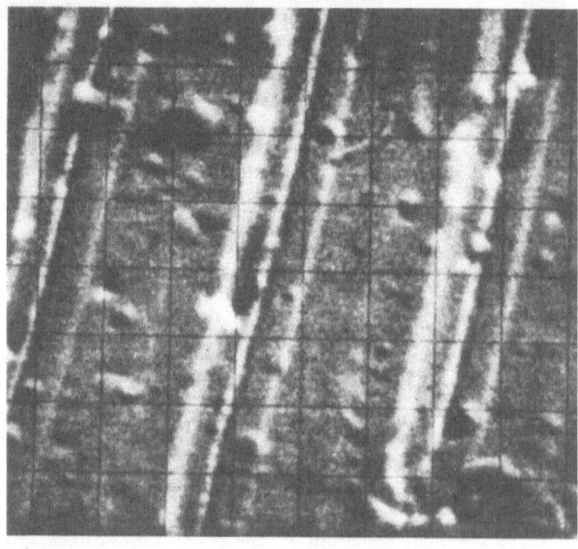

b

Fig. 7. (a) Backscattered primary electron scanning micrograph of a 1.72 μ d-spacing diffraction grating carbon replica
coated with uranium. (b) Same area of replica as it appears
using the secondary backscattered electrons.

electron beam of small cross section and high intensity, the probe forming lens for scanning electron microscopy should be capable of forming beams smaller than 0.1 μ in order to view the specimen at higher resolution. In addition, it must incorporate a number of exit windows for the emitted X-rays, a second means for viewing the specimen—a high quality light microscope—and a precision mechanical stage to move the specimen in the X–Y direction at micron intervals. Rotation and Z-axis movement to position the specimen plane in the Rowland circle of the X-ray spectrometers are also needed.

An instrument meeting most of these requirements was constructed by the author in 1961 ([8]). Figure 4 shows a close-up view of the light microscope attached to the specimen chamber for viewing the specimen either from below or from the opposite side of the electron beam. Micrometer X–Y screw-heads, the specimen chamber port for inserting the specimen, and above it the final objective lens are also visible. The complete electron optical column with the operator positioning the specimen under the electron beam is illustrated in Fig. 5.

Resolutions just below 1 μ were possible with this instrument when suitable prepared specimens were analyzed. Recent advances in instrumentation have considerably improved this resolution as well as the ease of operation and methods of data collection. A modern instrument* capable of less than 0.1 μ spatial resolution is shown in Fig. 6. This equipment is exclusively used for analysis of biological specimens at the Biodynamics Research Corporation laboratories. This electron-probe system is equipped with both secondary and primary backscattered detectors, and a transmission and reflex refractive light microscope for observing and positioning the specimen during analysis. It has X-ray detectors for the three spectrometers capable of analysis for elements with atomic numbers 4 through 92 and the associated electronic components as listed in the same figure.

Results obtained from test specimens to establish the resolution limits of the instrument are shown in Fig. 7 through 9. Figures 7a and 7b scanning electron micrographs utilizing backscattered electrons and secondary electrons respectively illustrate that resolutions less than 0.1 μ can be obtained. The specimen is a carbon–uranium coated diffraction grating replica of 1.72 μ d-spacing. The primary backscattered electron micrograph reveals more surface detail in depth without appreciable sacrifice in resolution than that observed in the secondary electron micrograph of exactly the same area. Since most biological problems of interest in our laboratory involve the examination of thin tissue sections or single cells, the electron-scanning resolution of this instrument was tested on frozen intestinal

*Materials Analysis Company, Palo Alto, Calif.

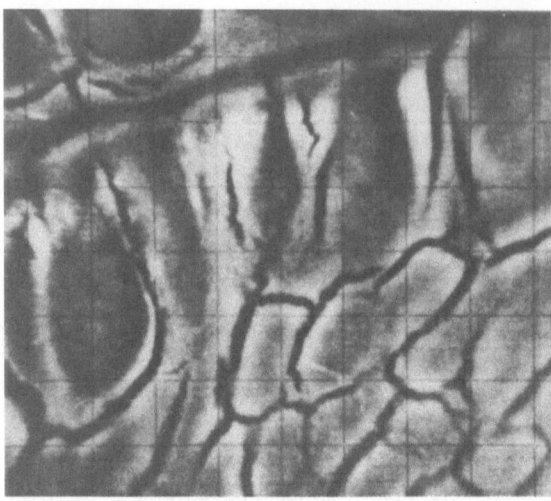

Fig. 8. Backscattered primary electron micrograph of 5-μ thick frozen section of amphibian small intestine.

tissue sections air dried at room temperature, as shown in Fig. 8. Although similar types of preparation can be used in light microscopy, from the backscattered scanning electron micrograph it is obvious that the structural detail is lost during the drying of the specimen and one could not proceed with further electron probe analysis.

3. ELECTRON BOMBARDMENT EFFECTS

3.1. Specimen Contamination

In electron microscopes, electron probes, scanning electron microscopes, and other similar instruments a "carbonaceous" layer is formed on the surface bombarded by electrons. According to Ennos [9] organic molecules from vacuum pump oil, high-vacuum greases, rubber gaskets, and other sources form a molecular layer on the specimen surface. The material is stabilized by polymerization when exposed to the electron beam. Subsequent layers continue to form as long as the area remains under the beam. Such a contamination layer can be seen as the dense line in Fig. 2. It formed on the bone section as the specimen was traversed under a stationary electron beam. In future instrument design, the use of vacuum systems employing ion pumps, vac-absorbers and different types of vacuum seals, as well as cold traps and rings placed in the vicinity of the specimen, might do much to minimize contamination.

3.2. Specimen Damage

Electron probe X-ray microanalysis is restricted to the identification and measurement of one or several elements present within the volume of specimen exposed to the electron beam. If changes in structure and elemental composition were brought by the impinging electrons, this would render the method of analysis inapplicable. No such changes, for elements with atomic number higher than 13, on properly prepared specimens, have been observed ([10]).

4. SPECIMEN PREPARATION

General procedures used for electron probe X-ray microanalysis in most studies to be reported in the symposium are schematically summarized in Fig. 9. Sections $0.2\ \mu$–$1\ \mu$ can be placed on 80–200 mesh screens coated with plastic or carbon films, or in some cases on a grid without any supporting film. In this type of preparation care must be taken to avoid interference of the supporting screen. Sections of similar or greater thickness (up to $0.5\ \mu$) can be placed on a highly polished surface such as high-purity quartz glass ($99.9\%\ SiO_2$), aluminum, or carbon. In our experience the best supporting surface for biological specimens from all considerations has been quartz, especially when the transmission light microscope is used. Tissue sections are pressed flat and carbon coated on the supporting surface. The specimen is rotated and tilted so that all exposed surfaces receive the same thickness of carbon (not to exceed 200 Å). This carbon film thickness has been found to provide adequate thermoelectric conductivity for specimen currents up to $0.5\ \mu A$. The experimental requirements determine whether, prior to carbon coating, the specimen is fixed, frozen dried or embedded before sectioning. Figures 10a and 10b show low-magnification reflex photomicrographs of 1-μ thick sections of epiphyseal plate cartilage (upper portion of micrograph

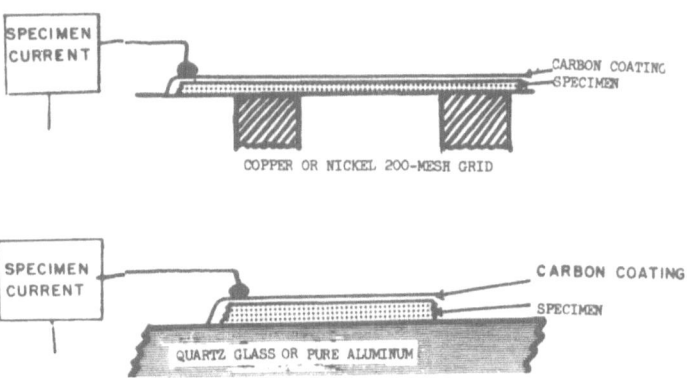

Fig. 9. Schematic, showing two methods of biological specimen preparation for electron probe analysis.

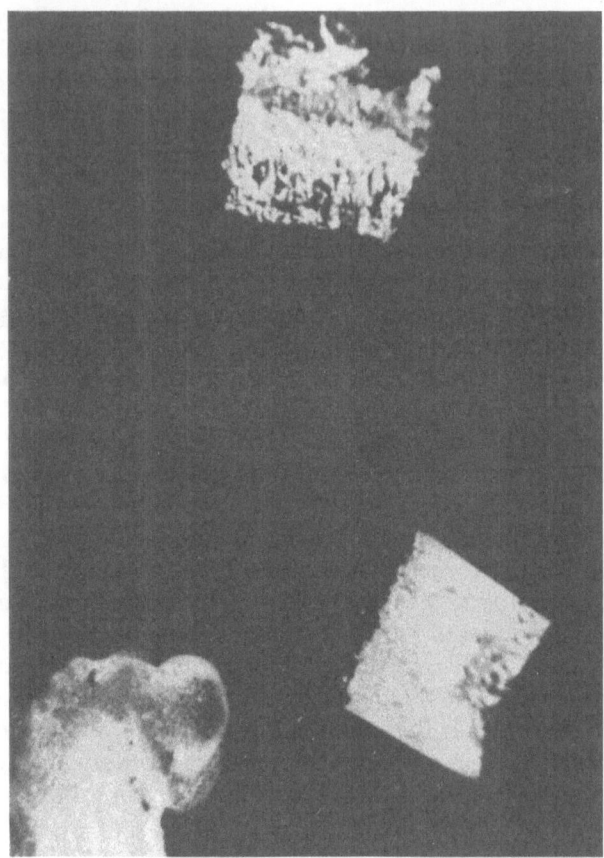

Fig. 10. (a) Low-magnification light micrograph showing a
thin section of the epiphyseal plate of the rat tibia (upper
portion of photo) and a similar thickness section of fluorapatite
standard seen at lower portion of micrograph.

10a). The aldehyde-fixed specimen was methacrylate embedded, and the
embedding medium was removed with xylene after the section was placed
on the quartz. In the lower left corner of Fig. 10a is a 1-μ thick section of
fluorapatite ground to submicron size particles embedded in gelatin and
then epoxy resin before sectioning. The latter section served as a standard
for calcium-phosphorus measurements made on the epiphyseal cartilage.

Figure 10b shows a low magnification (100 ×) backscattered electron
scanning micrograph of small intestinal epithelial cells embedded in epoxy
after fixation and staining in OsO_4 and cut at 1 μ. Some tissue detail is

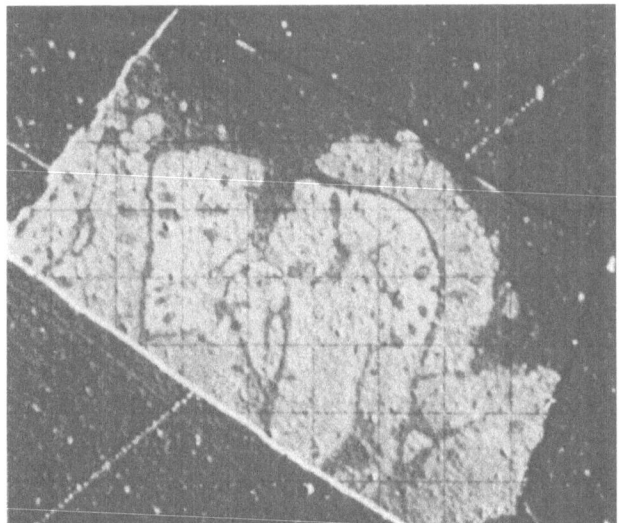

Fig. 10. (b) Backscattered primary electron micrograph of epoxy-embedded small intestine fixed in OsO_4 and then placed on a quartz surface.

discernible even at this low magnification. In addition, the transparency of the irregular polished quartz glass is apparent through the epoxy embedding medium.

5. APPLICATIONS

The biological applications of the electron probe analysis procedures, as noted from the program of this symposium, range from studies in bone mineralization (both normal and pathologic), and intracellular inclusions to electrolyte and amino acid transport studies across epithelial linings. Many laboratories throughout the world have initiated studies utilizing the electron probe that encompass various soft tissues of animal or plant origin.

5.1 Epiphyseal Cartilage

To know precisely the structure analyzed with the electron probe is one of the basic requirements of its analytical use. It is, therefore, important to the investigator for the interpretation of his results to observe the tissue section prepared for the probe with as many microscopic means as possible. To this end, the most logical starting point is light microscopy, as shown in

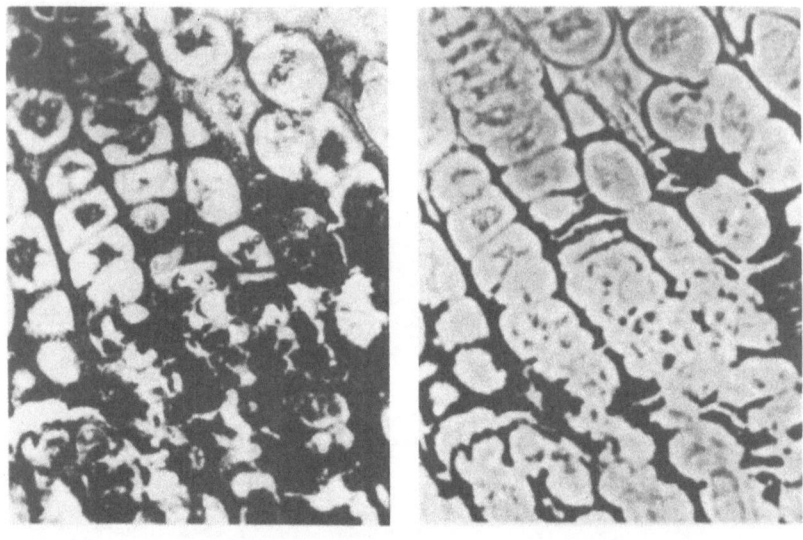

LIGHT MICROGRAPH   ELECTRON, S.C.

Fig. 11. Light micrograph and scanning electron micrograph made by using the electric current generated by the specimen as an electron beam was scanned over the same area of the sample as that shown in the light micrograph.

Fig. 11. The light micrograph of the 2-μ thick section shows the calcification zone, the disintegrating chondrocytes in rows, and the invading capillaries at the lower part of the photograph. The scanning electron micrograph of the same area was produced by using the electric current generated within the specimen as the electron beam was passed over the section. The calcifying matrix of the cartilage appears denser than the cellular components and their immediate environment.

Considerable improvement in spatial resolution is obtained from the same type of preparation using the backscattered primary electrons, as the micrographs of Fig. 12 (a–d) show. The area outlined by the square in the low magnification scanning micrograph (a) depicting the entire epiphyseal plate of the rat tibia can be seen at higher magnification in the remaining micrographs of this figure. Cellular detail, surfaces, and matrix structure are clearly illustrated.

This type of scanning electron presentation makes visible the structures for corollary X-ray analysis. Elements both of the ionic components (K, Cl, Na) and non-ionic (C, O, P, Ca, S), have been localized with great accuracy.

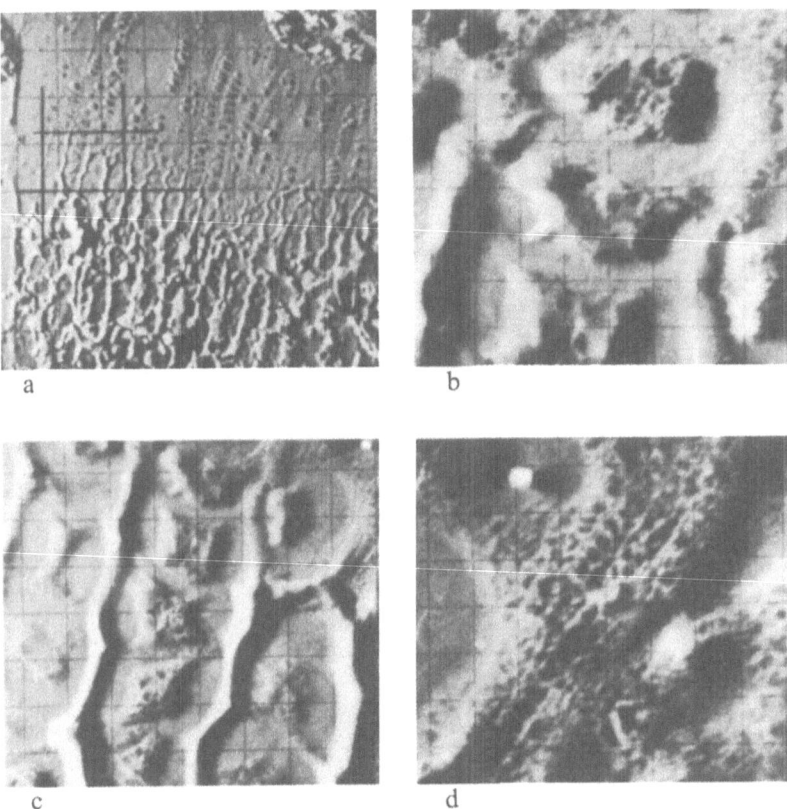

Fig. 12. (a) Low-magnification primary backscattered electron micrograph of a similarly prepared section as that shown in Fig. 11. The area shown in (c) is outlined by the square in (a) while (b) and (d) are enlargements of the same area depicting chondrocytes, their lacunar spaces and cartilage matrix.

5.2. Electrolytes in Cells

As my concluding example of the applicability of this instrument, scanning electron micrographs and X-ray intensities of the common and well-known electrolytes (K, Na, and Cl) of the red blood cells are presented in Figs. 13 and 14. A large area of an air dried preparation of amphibian (*R. pipiens*) nucleated red blood cells can be seen in Fig. 13a. One single cell with a collapsed surface and its nucleus protruding is illustrated in Fig. 13b. Figure 14 is a similar type preparation where, at the upper left, the specimen current scanning micrograph of a nucleated erythrocyte is shown.

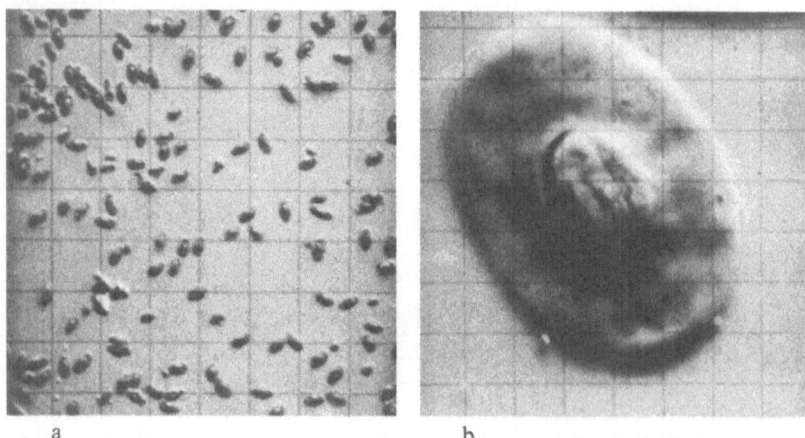

a   b

Fig. 13. (a) Survey primary backscattered electron micrograph of amphibian red blood cells. (b) Primary backscattered scanning electron micrograph of a single red blood cell (*R. pipiens*) air dried on a quartz surface.

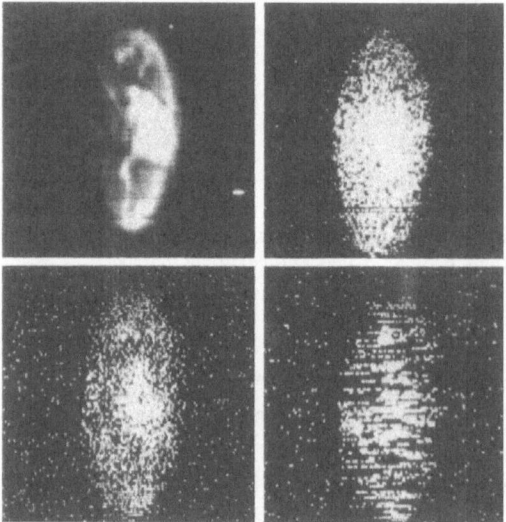

Fig. 14. The micrograph of a similar cell shown in Fig. 13 as seen via the specimen current, and clockwise the potassium, chlorine and sodium in the same cell.

The other three pictures clockwise are X-ray presentations showing potassium, chlorine and sodium distributions, respectively, of the same cell. Variation in X-ray intensities for these elements are seen within each cell. While this could reflect the variations in their concentration, this specimen was not prepared for the quantitative analysis of these elements. The mass thickness variations, surface irregularities and other changes introduced during preparation of the specimen could greatly distort the quantitative analysis. Uniformity of specimen mass thickness is of the utmost importance and must be monitored at all times during quantitative analysis.

REFERENCES

1. H. G. Moseley, The High Frequency Spectra of Elements, Part II, *Phil. Mag.* 27, 703 (1914).
2. J. Hillier, Electron Probe Analysis Employing X-ray Spectrography, U.S. Patent No. 2,418,029 (1947).
3. R. Castaing, Doctoral Thesis, University of Paris (1951).
4. I. B. Borovskii, Symposium on Problems in Metallurgy, Academy of Sciences of U.S.S.R., Moscow (1953).
5. L. S. Birks, and E. J. Brooks, Electron Probe X-ray Microanalyzer, *Rev. Sci. Instr.* 28, 709 (1957).
6. A. J. Tousimis, L. S. Birks, and E. J. Brooks, unpublished results (1957).
7. E. J. Brooks, A. J. Tousimis, and L. S. Birks, The Distribution of Calcium in the Epiphyseal Cartilage of the Rat Tibia Measured With the Electron Probe X-ray Microanalyzer, *J. Ultrastr. Res.* 7, 56 (1962).
8. A. J. Tousimis, Doctoral Thesis, George Washington University, Washington, D.C. (1963).
9. A. E. Ennos, The Sources of Electron Induced Contamination in Kinetic Vacuum Systems, *Brit. J. Appl. Phys.* 5, 27 (1954).
10. A. J. Tousimis, Electron Probe Microanalysis of Biological Specimens, in *X-ray Optics and X-ray Microanalysis* (H. H. Pattee, V. E. Cosslett, and Arne Engstrom, eds.) Academic Press, New York, pp. 539–557 (1963).

INDEX